Preface

Kermit Sigmon, author of the MATLAB Primer, passed away in January 1997. Kermit was a friend, colleague, and fellow avid bicyclist (although I'm a mere 10-mile-a-day commuter) with whom I shared an appreciation for the contribution that MATLAB has made to the mathematics, engineering, and scientific community. MATLAB is a powerful tool, and my hope is that in revising Kermit's book for MATLAB 6.1, you will be able to learn how to apply it to solving your own challenging problems in mathematics, science, and engineering.

A team at The MathWorks, Inc., revised the Fifth Edition. The current edition has undergone five major changes since the Fifth Edition, in addition to many smaller refinements. Only one of the five major changes was motivated by the release of MATLAB 6.1:

1. *Life is too short to spend writing DO loops.*[1] Over-using loops in MATLAB is a common mistake that new users make. To take full advantage of MATLAB's power, the emphasis on matrix operations has been strengthened, and the presentation of loops now appears after submatrices, colon notation, and matrix functions. A new section on the find function has been added. Many computations that would require nested loops with if statements in C, FORTRAN, or Java can be written as single loop-free

---

[1] John Little, co-founder of The MathWorks, Inc.

MATLAB statements with `find`. Avoiding loops makes your code faster and often easier to read.

2. In the Fifth Edition, the reader was often asked to come up with an appropriate matrix with which to try the examples. All examples are now fully described.

3. MATLAB 6.1 has a new and extensive graphical user interface, the MATLAB Desktop Environment.[2] Chapter 2, new to this edition, gives you an overview of all but two of MATLAB's primary windows (the other two are discussed later). Managing files and directories, starting MATLAB demos, getting help, command editing, debugging, and the like are explained in the new graphical user interface. This book was written for Release R12.1 (MATLAB Version 6.1 and the Symbolic Math Toolbox Version 2.1.2).

4. A new chapter on how to call a C routine from MATLAB has been added.

5. Sparse matrix ordering and visualization has been added to Chapter 13. Large matrices that arise in practical applications often have many zero entries. Taking advantage of sparsity allows you to solve problems in MATLAB that would otherwise be intractable.

I would like to thank Bob Stern, executive editor in Mathematics and Engineering at CRC Press, for giving

---

[2] Note that the Desktop Environment in Release R12.1 is not supported on HP and IBM Unix platforms.

Boca Raton   London   New York   Washington, D.C.

WITHDRAWN

The front cover shows a smooth free-form surface consisting of trimmed bicubic splines. The back cover shows a Bezier patch with its control polyhedron. The figures are courtesy of Jörg Peters and David Lutterkort, CISE Department, University of Florida. MATLAB code to generate the figures can be obtained from http://www.cise.ufl.edu/research/SurfLab.

MATLAB, Simulink, and Handle Graphics are registered trademarks of The MathWorks, Inc.

Library of Congress Cataloging-in-Publication Data

Sigmon, Kermit.
    MATLAB primer. — 6th ed. / Kermit Sigmon, Timothy A. Davis.
       p. cm.
    Rev. ed. of: MATLAB primer. 5th ed. / [MathWorks, Inc.] . c1998.
    Includes bibliographical references and index.
    ISBN 1-58488-294-8 (alk. paper)
    1. MATLAB. 2. Numerical analysis—Data processing. I. Davis,
Timothy A. II. MATLAB primer. III. Title.
    QA297 .S4787 2001
    519.4′0285′53042—dc21   2001047392

Visit the CRC Press Web site at www.crcpress.com

© 2002 by CRC Press LLC

No claim to original U.S. Government works
International Standard Book Number 1-58488-294-8
Library of Congress Card Number 2001047392
Printed in the United States of America   2 3 4 5 6 7 8 9 0
Printed on acid-free paper

me the opportunity to contribute to Kermit Sigmon's work. I would also like to thank Jörg Peters and David Lutterkort for providing the cover art. I would like to thank Naomi Fernandes, Madeline Leigh, Pei Li Li, Cleve Moler, Jim Tung, and Dave Wilson for their helpful comments on a draft of this book. Finally, I would like to thank The MathWorks, Inc., for providing software and technical support that assisted in the writing of this book.

Tim Davis

Associate Professor, Department of Computer and Information Science and Engineering
University of Florida
http://www.cise.ufl.edu/research/sparse

Introduction

MATLAB, developed by The MathWorks, Inc., integrates computation, visualization, and programming in a flexible, open environment. It offers engineers, scientists, and mathematicians an intuitive language for expressing problems and their solutions mathematically and graphically. Complex numeric and symbolic problems can be solved in a fraction of the time required with a programming language such as C, FORTRAN, or Java.

How to use this book: The purpose of this Primer is to help you begin to use MATLAB. It is not intended to be a substitute for the online help facility or the MATLAB documentation (such as *Getting Started with MATLAB* and *Using MATLAB*, available in printed form and online). The Primer can best be used hands-on. You are encouraged to work at the computer as you read the Primer and freely experiment with the examples. This Primer, along with the online help facility, usually suffices for students in a class requiring the use of MATLAB.

Start with the examples at the beginning of each chapter. In this way, you will create all of the matrices and M-files used in the examples (with one exception: an M-file you write in Chapter 7 is used in later chapters).

Larger examples (M-files and MEX-files) are on the web at http://www.cise.ufl.edu/research/sparse/MATLAB and http://www.crcpress.com.

Pull-down menu selections are described using the following style. Selecting the View menu, and then the

Desktop Layout submenu, and then the Simple menu
item is written as View ▶ Desktop Layout ▶ Simple.

You should liberally use the online help facility for more
detailed information. Selecting Help ▶ MATLAB Help
brings up the Help window. You can also type help in
the Command window. See Sections 2.1 or 15.1 for more
information.

How to obtain MATLAB: Version 6.1 of MATLAB is
available for Unix (Sun, HP, Compaq Alpha, IBM,
Silicon Graphics, and Linux), and Microsoft Windows.
MATLAB 5 is also available for the Apple Macintosh. A
Student Version of MATLAB is available from The
MathWorks, Inc., for Microsoft Windows and Linux; it
includes MATLAB, Simulink, and key functions of the
Symbolic Math Toolbox. Everything discussed in this
book can be done in the Student Version of MATLAB,
with the exception of advanced features of the Symbolic
Math Toolbox discussed in Section 14.11. The Student
Edition of MATLAB Version 5, from Prentice-Hall, was
limited in the size of the matrices it could operate on.
These restrictions have been removed in the Student
Version of MATLAB Versions 6 and 6.1. For more
information on MATLAB, contact:

The MathWorks, Inc.
3 Apple Hill Drive
Natick, MA, 01760-2098 USA
Phone: 508–647–7000
Fax: 508–647–7101
Email: info@mathworks.com
Web: http://www.mathworks.com

Table of Contents

1. Accessing MATLAB

On Unix systems you can enter MATLAB with the system command `matlab` and exit MATLAB with the MATLAB command `quit` or `exit`. In Microsoft Windows, the Apple Macintosh, and in some Unix window systems, just double-click on the MATLAB icon:

MATLAB 6.1

2. The MATLAB Desktop

MATLAB has an extensive graphical user interface. When MATLAB starts, the MATLAB window will appear, with several subwindows and menu bars.

All of MATLAB's windows are docked, which means that they are tiled on the main MATLAB window. You can undock a window by clicking its undock button:

Dock it with View ▶ Dock. Close a window by clicking its close button:

Reshape the window tiling by clicking on and dragging the window edges.

The menu bar at the top of the MATLAB window contains a set of buttons and pull-down menus for working with M-files, windows, preferences and other settings, web resources for MATLAB, and online MATLAB help. For example, if you prefer a simpler font than the default one, select File ▶ Preferences, click on ⊞ General and then Font & Colors. Select Lucida Console (on a PC) or DialogInput (on Unix) in place of the default Monospaced font, and click OK.

2.1 Help window

This window is the most useful window for beginning MATLAB users. Select Help ▶ MATLAB Help. The Help window has most of the features you would see in any web browser (clickable links, a back button, and a search engine, for example). The Help Navigator on the left shows where you are in the MATLAB online documentation. I'll refer to the online Help sections as Help: MATLAB: Getting Started: Introduction, for example. Click on MATLAB in the Help Navigator, and you'll see the MATLAB Roadmap (or Help: MATLAB for short). Printable versions of the documentation are also available (see Help: MATLAB: Printable Documentation (PDF)).

You can also use the help command, typed in the Command window. For example, the command help eig will give information about the eigenvalue function eig. See the list of functions in the last section of this Primer for a brief summary of help for a function. You can also preview some of the features of MATLAB by first entering the command demo or by selecting Help ▶ Demos, and then selecting from the options offered.

2.2 Launch Pad window

This allows you to start up demos and other windows not present when you start MATLAB. Try Launch Pad: MATLAB: Demos and run one of the demos from the MATLAB Demo window.

2.3 Command window

MATLAB expressions and statements are evaluated as you type them in the Command window, and results of the computation are displayed there too. Expressions and statements are also used in M-files (more on this in Chapter 7). They are usually of the form:

> *variable = expression*

or simply:

> *expression*

Expressions are usually composed from operators, functions, and variable names. Evaluation of the expression produces a matrix (or other data type), which is then displayed on the screen or assigned to a variable for future use. If the variable name and = sign are omitted, a variable ans (for answer) is automatically created to which the result is assigned.

A statement is normally terminated with the carriage return. However, a statement can be continued to the next line with three periods (. . .) followed by a carriage return. On the other hand, several statements can be placed on a single line separated by commas or semicolons. If the last character of a statement is a semicolon, display of the result is suppressed, but the assignment is carried out. This is essential in suppressing unwanted display of intermediate results.

Click on the Workspace tab to bring up the Workspace window (it starts out underneath the Launch Pad window) so you can see a list of the variables you create, and type this command in the Command window:

```
A = [1 2 3 ; 4 5 6 ; -1 7 9]
```

or this one:

```
A = [
1 2 3
4 5 6
-1 7 9]
```

in the Command window. Either one creates the obvious 3-by-3 matrix and assigns it to a variable A. Try it. You will see the array A in your Workspace window. MATLAB is case-sensitive in the names of commands, functions, and variables, so A and a are two different variables. A comma or blank separates the elements within a row of a matrix (sometimes a comma is necessary to split the expressions, because a blank can be ambiguous). A semicolon ends a row. When listing a number in exponential form (e.g., 2.34e-9), blank spaces must be avoided. Matrices can also be constructed from other matrices. If A is the 3-by-3 matrix shown above, then:

```
C = [A, A' ; [12 13 14], (zeros (1,3))]
```

creates a 4-by-6 matrix. Try it to see what C is. The quote mark in A' means the transpose of A. Be sure to use the correct single quote mark (just to the left of the enter or return key on most keyboards). Parentheses are needed around expressions if they would otherwise be ambiguous. If you leave out the parentheses around (zeros(1,3)), you will get an error message. The zeros function is described in Section 5.1.

When you typed the last two commands, the matrices A and C were created and displayed in the Workspace window.

You can save the Command window dialog with the diary command:

```
diary filename
```

This causes what appears subsequently on the screen (except graphics) to be written to the named file (if the *filename* is omitted, it is written to a default file named diary) until you type the command diary off; the command diary on causes writing to the file to resume. When finished, you can edit the file as desired and print it out. For hard copy of graphics, see Section 10.10.

The command line in MATLAB can be easily edited in the Command window. The cursor can be positioned with the left and right arrows and the Backspace (or Delete) key used to delete the character to the left of the cursor. Type help cedit to see more command-line editing features.

A convenient feature is use of the up and down arrows to scroll through the stack of previous commands. You can, therefore, recall a previous command line, edit it, and execute the revised line. Try this by first modifying the matrix A by adding one to each of its elements:

```
A = A + 1
```

You can change C to reflect this change in A by retyping the lengthy command C = ... above, but it is easier to hit the up arrow key until you see the command you want, and then hit enter.

You can clear the Command window with the clc command or with Edit ▶ Clear Command window.

Although all numeric computations in MATLAB are performed with about 16 decimal digits of precision, the format of the displayed output can be controlled by the following commands:

format short	fixed point, 5 digits
format long	fixed point, 15 digits
format short e	scientific notation, 5 digits
format long e	scientific notation, 15 digits
format short g	fixed or floating-point, 5 digits
format long g	fixed or floating-point, 15 digits
format hex	hexadecimal format
format +	+, -, and blank
format bank	dollars and cents
format rat	approximate ratio of small integers

format short is the default. Once invoked, the chosen format remains in effect until changed. These commands only modify the display, not the precision of the number.

The command format compact suppresses most blank lines, allowing more information to be placed on the screen or page. The command format loose returns to

6

the non-compact format. These two commands are independent of the other format commands.

You can pause the output in the Command window with the `more on` command. Type `more off` to turn this feature off.

2.4 Workspace window

This lists variables that you have either entered or computed in your MATLAB session.

There are many fundamental data types (or classes) in MATLAB, each one a multidimensional array. The classes that we will concern ourselves with most are rectangular numerical arrays with possibly complex entries, and possibly sparse. An array of this type is called a matrix. A matrix with only one row or one column is called a vector (row vectors and column vectors behave differently; they are more than mere one-dimensional arrays). A 1–by–1 matrix is called a scalar.

Arrays can be introduced into MATLAB in several different ways. They can be entered as an explicit list of elements (as you did for matrix A), generated by statements and functions (as you did for matrix C), created in a file with your favorite text editor, or loaded from external data files or applications (see `Help: MATLAB: Getting Started: Manipulating Matrices`). You can also write your own functions (M-files, or mexFunctions in C, FORTRAN, or Java) that create and operate on matrices. All the matrices and other variables that you create, except those internal to M-files (see Chapter 7), are shown in your Workspace window.

7

The command who (or whos) lists the variables currently in the workspace. Try typing whos; you should see a list of variables including A and C, with their type and size. A variable or function can be cleared from the workspace with the command clear *variablename* or by right-clicking the variable in the Workspace editor and selecting Delete Selection. The command clear alone clears all non-permanent variables.

When you log out or exit MATLAB, all variables are lost. However, invoking the command save before exiting causes all variables to be written to a machine-readable file named matlab.mat. When you later reenter MATLAB, the command load will restore the workspace to its former state. Commands save and load take file names and variable names as optional arguments (type help save and help load). Try typing the commands save, clear, and then load, and watch what happens after each command.

2.5 Command History window

This window lists the commands typed in so far. You can re-execute a command from this window by double-clicking or dragging the command into the Command window. Try double-clicking on the command:

```
A = A + 1
```

shown in your Command History window. For more options, right-click on a line of the Command window.

2.6 Array Editor window

Once an array exists, it can be modified with the Array Editor, which acts like a spreadsheet for matrices. Go to the Workspace window and double-click on the matrix C. Click on an entry in C and change it, and try changing the size of C. Go back to the Command window and type:

```
C
```

and you will see your new array C. You can also edit the matrix C by typing the command `openvar('C')`.

2.7 Current Directory window

Your current directory is where MATLAB looks for your M-files (see Chapter 10), and for workspace (`.mat`) files that you `load` and `save`. You can also load and save matrices as ASCII files and edit them with your favorite text editor. The file should consist of a rectangular array of just the numeric matrix entries. Use a text editor to create a file in your current directory called `mymatrix.txt` that contains these 2 lines:

```
22 67
12 33
```

Type the command `load mymatrix.txt`, and the file will be loaded from the current directory to the variable `mymatrix`. The file extension (`.txt` in this example) can be anything except `.mat`. Large matrices may also be entered with an M-file (see Section 7.7).

You can use the menus and buttons in the Current Directory window to peruse your files, or you can use commands typed in the Command window. The command `pwd` returns the name of the current directory, and `cd` will change the current directory. The command `dir` lists the contents of the working directory, whereas the command `what` lists only the MATLAB-specific files in the directory, grouped by file type. The MATLAB commands `delete` and `type` can be used to delete a file and display an M-file in the Command window, respectively.

2.8 MATLAB's path

M-files must be in a directory accessible to MATLAB. M-files in the current directory are always accessible. The current list of directories in MATLAB's search path is obtained by the command `path`. This command can also be used to add or delete directories from the search path. See `help path`. The command `which` locates functions and files on the path. For example, type `which hilb`. You can modify your MATLAB path with the command `path`, or `pathtool`, which brings up another window. You can also select `File ▶ Set Path`.

3. Matrices and Matrix Operations

You have now seen most of MATLAB's windows and what they can do. Now take a look at how you can use MATLAB to work on matrices and other data types.

3.1 Referencing individual entries

Individual matrix and vector entries can be referenced with indices inside parentheses. For example, `A(2,3)` denotes the entry in the second row, third column of matrix A. Try:

```
A = [1 2 3 ; 4 5 6 ; -1 7 9]
A (2,3)
```

Next, create a column vector, x, with:

```
x = [3 2 1]'
```

or equivalently:

```
x = [3 ; 2 ; 1]
```

With this vector, x(3) denotes the third coordinate of vector x, with a value of 1. Higher dimensional arrays are similarly indexed. A matrix or a vector accepts only positive integers as indices.

A two-dimensional array can be indexed as if it were a one-dimensional vector. If A is m-by-n, then A(i,j) is the same as A(i+(j-1)*m). This feature is most often used with the find function (see Section 5.5).

3.2 Matrix operators

The following matrix operators are available in MATLAB:

+	addition
–	subtraction or negation
*	multiplication
^	power
'	transpose (real) or conjugate transpose (complex)
.'	transpose (real or complex)
\	left division
/	right division

These matrix operators apply, of course, to scalars (1-by-1 matrices) as well. If the sizes of the matrices are incompatible for the matrix operation, an error message will result, except in the case of scalar-matrix operations (for addition, subtraction, division, and multiplication, in which case each entry of the matrix is operated on by the scalar, as in A=A+1). Also try the commands:

```
A^2
A*x
```

If x and y are both column vectors, then x'*y is their inner (or dot) product, and x*y' is their outer (or cross) product. Try these commands:

```
y = [1 2 3]'
x'*y
x*y'
```

3.3 Matrix division

The matrix division operations deserve special comment. If A is an invertible square matrix and b is a compatible column vector, or respectively a compatible row vector, then x=A\b is the solution of A*x=b, and x=b/A is the solution of x*A=b. If A is square and non-singular, then A\b and b/A are mathematically the same as inv(A)*b and b*inv(A), respectively, where inv(A) computes the inverse of A. The left and right division operators are more accurate and efficient. In left division, if A is square, then it is factored using Gaussian elimination, and these factors are used to solve A*x=b. If A is not square, the under- or over-determined system is solved in the least squares sense. Right division is defined in terms of left division by b/A = (A'\b')'. Try this:

```
A = [1 2 ; 3 4]
b = [4 10]'
x = A\b
```

The solution to A*x=b is the column vector x=[2 ; 1].

3.4 Entry-wise operators

Matrix addition and subtraction already operate entry-wise, but the other matrix operations do not. These other operators (*, ∧, \, and /) can be made to operate entry-wise by preceding them by a period. For example, either:

```
[1 2 3 4] .* [1 2 3 4]
```

or:

```
[1 2 3 4] .∧ 2
```

will yield [1 4 9 16]. Try it. This is particularly useful when using MATLAB graphics.

Also compare A∧2 with A.∧2.

3.5 Relational operators

The relational operators in MATLAB are:

<	less than
>	greater than
<=	less than or equal
>=	greater than or equal
==	equal
~=	not equal

They all operate entry-wise. Note that = is used in an assignment statement whereas == is a relational operator. Relational operators may be connected by logical operators:

&	and
|	or
~	not

When applied to scalars, the result is 1 or 0 depending on whether the expression is true or false. Try entering 3 < 5, 3 > 5, 3 == 5, and 3 == 3. When applied to matrices of the same size, the result is a matrix of ones and zeros giving the value of the expression between corresponding entries. You can also compare elements of a matrix with a scalar. Try:

```
A = [1 2 ; 3 4]
A >= 2
B = [1 3 ; 4 2]
A < B
```

In logical expressions, a nonzero value is interpreted as true, and a zero is interpreted as false. Thus, ~0 is 1, ~3 is 0, and 4 & 5 is 1, for example.

3.6 Complex numbers

MATLAB allows complex numbers in most of its operations and functions. Two convenient ways to enter complex matrices are:

```
B = [1 2 ; 3 4] + i*[5 6 ; 7 8]
B = [1+5i, 2+6i ; 3+7i, 4+8i]
```

Either i or j may be used as the imaginary unit. If, however, you use i and j as variables and overwrite their values, you may generate a new imaginary unit with, say, ii=sqrt(-1). You can also use 1i or 1j, which cannot be reassigned and are always equal to the imaginary unit. Thus,

```
B = [1 2 ; 3 4] + 1i*[5 6 ; 7 8]
```

14

generates the same matrix B, even if i has been reassigned. See Section 8.2 to find out if i has been reassigned.

3.7 Strings

Enclosing text in single quotes forms strings with the char data type:

```
S = 'I love MATLAB'
```

To include a single quote inside a string, use two of them together, as in:

```
S = 'Green''s function'
```

Strings, numeric matrices, and other data types can be displayed with the function disp. Try disp(S) and disp(B).

3.8 Other data types

MATLAB supports many other data types, including sparse matrices, multidimensional arrays, cell arrays, and structures.

Sparse matrices are stored in a special way that does not require space for zero entries. MATLAB has efficient methods of operating on sparse matrices. Type help sparse, and help full, look in Help: MATLAB: Using MATLAB: Mathematics: Sparse Matrices, or see Chapter 13. Sparse matrices are allowed as arguments for most, but not all, MATLAB operators and functions where a normal matrix is allowed.

`D=zeros(3,5,4,2)` creates a 4-dimensional array of size 3-by-5-by-4-by-2. Multidimensional arrays may also be built up using `cat` (short for concatenation).

Cell arrays are collections of other arrays or variables of varying types and are formed using curly braces. For example,

```
c = {[3 2 1] ,'I love MATLAB'}
```

creates a cell array. The expression `c{1}` is a row vector of length 3, while `c{2}` is a string.

A `struct` is variable with one or more parts, each of which has its own type. Try, for example,

```
x.particle = 'electron'
x.position = [2 0 3]
x.spin = 'up'
```

The variable x describes an object with several characteristics, each with its own type.

You may create additional data objects and classes using overloading (see `help class`).

4. Submatrices and Colon Notation

Vectors and submatrices are often used in MATLAB to achieve fairly complex data manipulation effects. Colon notation (which is used to both generate vectors and reference submatrices) and subscripting by integral vectors are keys to efficient manipulation of these objects. Creative use of these features minimizes the use of loops (which slows MATLAB) and makes code simple and

readable. Special effort should be made to become
familiar with them.

4.1 Generating vectors

The expression 1:5 is the row vector [1 2 3 4 5].
The numbers need not be integers, and the increment need
not be one. For example, 0:0.2:1 gives [0 0.2 0.4
0.6 0.8 1], and 5:-1:1 gives [5 4 3 2 1]. These
vectors are commonly used in for loops, described in
Section 6.1. Be careful how you mix the colon operator
with other operators. Compare 1:5-3 with (1:5)-3.

4.2 Accessing submatrices

Colon notation can be used to access submatrices of a
matrix. To try this out, first type the two commands:

```
A = rand (6,6)
B = rand (6,4)
```

which generate a random 6-by-6 matrix A and a random
6-by-4 matrix B (see Section 5.1).

A(1:4,3) is the column vector consisting of the first
four entries of the third column of A.

A colon by itself denotes an entire row or column:
A(:,3) is the third column of A, and A(1:4,:) is the
first four rows.

Arbitrary integral vectors can be used as subscripts:
A(:,[2 4]) contains as columns, columns 2 and 4 of A.
Such subscripting can be used on both sides of an
assignment statement:

```
A (:,[2 4 5]) = B (:,1:3)
```

17

replaces columns 2, 4, 5 of A with the first three columns
of B. Try it. Note that the entire altered matrix A is
displayed and assigned.

Columns 2 and 4 of A can be multiplied on the right by
the 2-by-2 matrix [1 2 ; 3 4]:

```
A (:,[2 4]) = A (:,[2 4]) * [1 2 ; 3 4]
```

Once again, the entire altered matrix is displayed and
assigned. Submatrix operations are a convenient way to
perform many useful computations. For example, a
Givens rotation of rows 3 and 5 of the matrix A to zero
out the A(3,1) entry can be written as:

```
a = A (5,1)
b = A (3,1)
G = [a b ; -b a] / norm ([a b])
A ([5 3], :) = G * A ([5 3], :)
```

(assuming norm([a b]) is not zero). You can also
assign a scalar to all entries of a submatrix. Try:

```
A (:, [2 4]) = 99
```

You can delete rows or columns of a matrix by assigning
the empty matrix ([]) to them. Try:

```
A (:, [2 4]) = []
```

In an array index expression, end denotes the index of the
last element. Try:

```
x = rand (1,5)
x = x (end:-1:1)
```

To appreciate the usefulness of these features, compare these MATLAB statements with a C, FORTRAN, or Java routine to do the same operation.

5. MATLAB Functions

MATLAB has a wide assortment of built-in functions. You have already seen some of them, such as `zeros`, `rand`, and `inv`. This section describes the more common matrix manipulation functions. For a more complete list, see Chapter 14, or Help: MATLAB: Reference: MATLAB Function Reference.

5.1 Constructing matrices

Convenient matrix building functions are:

eye	identity matrix
zeros	matrix of zeros
ones	matrix of ones
diag	create or extract diagonals
triu	upper triangular part of a matrix
tril	lower triangular part of a matrix
rand	randomly generated matrix
hilb	Hilbert matrix
magic	magic square
toeplitz	Toeplitz matrix

The command `rand(n)` creates an n-by-n matrix with randomly generated entries distributed uniformly between 0 and 1 while `rand(m,n)` creates an m-by-n matrix (m and n denote, of course, positive integers). Try:

```
A = rand (3)
```

`rand('state',0)` resets the random number generator.
`zeros(m,n)` produces an m-by-n matrix of zeros, and
`zeros(n)` produces an n-by-n one. If A is a matrix, then
`zeros(size(A))` produces a matrix of zeros having the
same size as A. If x is a vector, `diag(x)` is the diagonal
matrix with x down the diagonal; if A is a matrix, then
`diag(A)` is a vector consisting of the diagonal of A. Try:

```
x = 1:3
diag (x)
diag (A)
diag (diag (A))
```

Matrices can be built from blocks. Try creating this 5-by-
5 matrix:

```
B = [A, (zeros (3,2)) ;
(pi * ones (2,3)), (eye (2))]
```

`magic(n)` creates an n-by-n matrix that is a magic
square (rows, columns, and diagonals have common
sum); `hilb(n)` creates the n-by-n Hilbert matrix, the
king of ill-conditioned matrices. Matrices can also be
generated with a `for` loop (see Section 6.1). `triu` and
`tril` extract upper and lower triangular parts of a matrix.
Try:

```
triu (A)
triu (A) == A
```

5.2 Scalar functions

Certain MATLAB functions operate essentially on scalars
but operate entry-wise when applied to a vector or matrix.
The most common such functions are:

```
abs    ceil   log     sign
acos   cos    log10   sin
```

```
asin    exp     rem     sqrt
atan    floor   round   tan
```

The following statements, for example, will generate a sine table. Try it.

```
x = (0:0.1:2)'
y = sin (x)
[x y]
```

Note that because sin operates entry-wise, it produces a vector y from the vector x.

5.3 Vector functions

Other MATLAB functions operate essentially on a vector (row or column) but act on an m-by-n matrix (m > 2) in a column-by-column fashion to produce a row vector containing the results of their application to each column. Row-by-row action can be obtained by using the transpose (mean(A')', for example) or by specifying the dimension along which to operate (mean(A,2), for example). A few of these functions are:

```
max    sum    median    any    sort
min    prod   mean      all    std
```

The maximum entry in a matrix A is given by max(max(A)) rather than max(A). Try it.

5.4 Matrix functions

Much of MATLAB's power comes from its matrix functions. The most useful ones are:

```
eig     eigenvalues and eigenvectors
chol    Cholesky factorization
svd     singular value decomposition
```

inv	inverse
lu	LU factorization
qr	QR factorization
hess	Hessenberg form
schur	Schur decomposition
rref	reduced row echelon form
expm	matrix exponential
sqrtm	matrix square root
poly	characteristic polynomial
det	determinant
size	size of an array
length	length of a vector
norm	1–norm, 2–norm, Frobenius–norm, ∞–norm
cond	condition number in the 2–norm
rank	rank
kron	Kronecker tensor product
find	find indices of nonzero entries

MATLAB functions may have single or multiple output arguments. For example,

```
y = eig (A)
```

produces a column vector containing the eigenvalues of A, whereas:

```
[U, D] = eig (A)
```

produces a matrix U whose columns are the eigenvectors of A and a diagonal matrix D with the eigenvalues of A on its diagonal. Try it.

5.5 The find function

The find function is unlike the others. find(x), where x is a vector, returns an array of indices of nonzero entries in x. This is often used in conjunction with relational operators. Suppose you want a vector y that consists of all the values in x greater than 1. Try:

```
x = 2*rand (1,5)
y = x (find (x > 1))
```

For matrices,

```
[i,j,x] = find (A)
```

returns three vectors, with one entry in i, j, and x for each nonzero in A (row index, column index, and numerical value, respectively). With this matrix A, try:

```
[i,j,x] = find (A > .5)
[i  j  x]
```

and you will see a list of pairs of row and column indices where A is greater than .5. However, x is a vector of values from the matrix expression A > .5, not from the matrix A. Getting the values of A that are larger than .5 without using a loop (see Section 6.1) requires one-dimensional array indexing. Try:

```
k = find (A > .5)
A (k)
A (k) = A (k) + 99
```

The loop-based analog of this computation is shown in Section 6.1.

Here's a more complex example. A square matrix A is diagonally dominant if

$$|a_{ii}| > \sum_{j \neq i} |a_{ij}| \qquad \text{for each row } i.$$

First, enter a matrix that is not diagonally dominant. Try:

```
A = [
-1  2  3 -4
 0  2 -1  0
 1  2  9  1
-3  4  1  1]
```

These statements compute a vector i containing indices of rows that violate diagonal dominance (rows 1 and 4 for this matrix A).

```
d = diag (A)
a = abs (d)
f = sum (abs (A), 2) - a
i = find (f >= a)
```

Next, modify the diagonal entries to make the matrix just barely diagonally dominant, while still preserving the sign of the diagonal:

```
[m n] = size (A)
k = i + (i-1)*m
tol = 100 * eps
s = 2 * (d (i) >= 0) - 1
A (k) = (1+tol) * s .* max (f (i), tol)
```

The variable eps (epsilon) gives the smallest value such that 1+eps > 1, about 10^{-16} on most computers. It is useful in specifying tolerances for convergence of iterative processes and in problems like this one. The

24

odd-looking statement that computes s is nearly the same
as s=sign(d(i)), except that here we want s to be one
when d(i) is zero. We'll come back to this diagonal
dominance problem later on.

6. Control Flow Statements

In their basic forms, these MATLAB flow control
statements operate like those in most computer languages.
Indenting the statements of a loop or conditional
statement is optional, but it helps readability to follow a
standard convention.

6.1 The for loop

This loop:

```
n = 10
x = []
for i = 1:n
    x = [x, i^2]
end
```

produces a vector of length 10, and

```
n = 10
x = []
for i = n:-1:1
    x = [x, i^2]
end
```

produces the same vector in reverse order. Try them.
The vector x grows in size at each iteration. Note that a
matrix may be empty (such as x=[]). The statements:

```
m = 6
n = 4
for i = 1:m
    for j = 1:n
```

```
                H (i,j) = 1/(i+j-1) ;
        end
    end
    H
```

produce and display in the Command window the 6-by-4
Hilbert matrix. The last H displays the final result. The
semicolon on the inner statement is essential to suppress
the display of unwanted intermediate results. If you leave
off the semicolon, you will see that H grows in size as the
computation proceeds. This can be slow if m and n are
large. It is more efficient to preallocate the matrix H with
the statement H=zeros(m,n) before computing it. Type
the command type hilb to see a more efficient way to
produce a square Hilbert matrix.

Here is the counterpart of the one-dimensional indexing
exercise from Section 5.5. It adds 99 to each entry of the
matrix that is larger than .5, using two for loops instead
of a single find. This method is much slower.

```
A = rand (3)
[m n] = size (A) ;
for j = 1:n
    for i = 1:m
        if (A (i,j) > .5)
            A (i,j) = A (i,j) + 99 ;
        end
    end
end
A
```

The for statement permits any matrix expression to be
used instead of 1:n. The index variable consecutively
assumes the value of each column of the expression. For
example,

```
s = 0 ;
for c = H
    s = s + sum (c) ;
end
```

computes the sum of all entries of the matrix H by adding its column sums (of course, sum(sum(H)) does it more efficiently; see Section 5.3). In fact, since $1:n = [1\ 2\ 3\ \ldots\ n]$, this column-by-column assignment is what occurs with for i = 1:n.

6.2 The while loop

The general form of a while loop is:

```
while expression
    statements
end
```

The *statements* will be repeatedly executed as long as the *expression* remains true. For example, for a given number a, the following computes and displays the smallest nonnegative integer n such that $2^n > a$:

```
a = 1e9
n = 0
while 2^n <= a
    n = n + 1 ;
end
n
```

Note that you can compute the same value n more efficiently by using the log2 function:

```
[f,n] = log2 (a)
```

You can terminate a for or while loop with the break statement and skip to the next iteration with the continue statement.

6.3 The if statement

The general form of a simple `if` statement is:

```
if expression
    statements
end
```

The *statements* will be executed only if the *expression* is true. Multiple conditions also possible:

```
for n = -2:5
    if n < 0
        parity = 0 ;
    elseif rem (n,2) == 0
        parity = 2 ;
    else
        parity = 1 ;
    end
    n
    parity
end
```

The `else` and `elseif` are optional. If the `else` part is used, it must come last.

6.4 The switch statement

The `switch` statement is just like the `if` statement. If you have one expression that you want to compare against several others, then a `switch` statement can be more concise than the corresponding `if` statement. See `help switch` for more information.

6.5 The try/catch statement

Matrix computations can fail because of characteristics of the matrices that are hard to determine before doing the computation. If the failure is severe, your script or

function (see Chapter 7) may be terminated. The
try/catch statement allows you to compute
optimistically and then recover if those computations fail.
The general form is:

```
try
    statements
catch
    statements
end
```

The first block of statements is executed. If an error
occurs, those statements are terminated, and the second
block of statements is executed. You cannot do this with
an if statement. See help try.

6.6 Matrix expressions (if and while)

A matrix expression is interpreted by if and while to be
true if **every** entry of the matrix expression is nonzero.
Enter these two matrices:

```
A = [ 1 2 ; 3 4 ]
B = [ 2 3 ; 3 5 ]
```

If you wish to execute a statement when matrices A and B
are equal, you could type:

```
if A == B
    disp ('A and B are equal')
end
```

If you wish to execute a statement when A and B are not
equal, you would type:

```
if any (any (A ~= B))
    disp ('A and B are not equal')
end
```

or, more simply,

```
if A == B else
    disp ('A and B are not equal')
end
```

Note that the seemingly obvious:

```
if A ~= B
    disp ('not what you think')
end
```

will not give what is intended because the statement
would execute only if each of the corresponding entries of
A and B differ. The functions any and all can be
creatively used to reduce matrix expressions to vectors or
scalars. Two anys are required above because any is a
vector operator (see Section 5.3). In logical terms, any
and all correspond to the existential ($\exists$) and universal
($\forall$) quantifiers, respectively, applied to each column of a
matrix or each entry of a row or column vector. Like most
vector functions, any and all can be applied to
dimensions of a matrix other than the columns.

Thus, an if statement with a two-dimensional matrix
expression is equivalent to:

```
if all (all (expression))
    statement
end
```

6.7 Infinite loops

With loops, it is possible to execute a command that will
never stop. Typing Ctrl-C stops a runaway display or
computation. Try:

```
i = 1
while i > 0
    i = i + 1
end
```

then type Ctrl-C to terminate this loop.

7. M-files

MATLAB can execute a sequence of statements stored in files. These are called M-files because they must have the file type .m as the last part of their filename.

7.1 M-file Editor/Debugger window

Much of your work with MATLAB will be in creating and refining M-files. M-files are usually created using your favorite text editor or with MATLAB's M-file Editor/Debugger. See also Help: MATLAB: Using MATLAB: Development Environment: Editing and Debugging M-Files.

There are two types of M-files: script files and function files. In this exercise, you will incrementally develop and debug a script and then a function for making a matrix diagonally dominant (see Section 5.5). Select File ▶ New ▶ M-file to start a new M-file, or click:

Type in these lines in the Editor,

```
f = sum (A, 2) ;
A = A + diag (f) ;
```

and save the file as **ddom.m** by clicking:

You've just created a MATLAB script file.[3] The
semicolons are there because you normally do not want to
see the results of every line of a script or function.

7.2 Script files

A script file consists of a sequence of normal MATLAB
statements. Typing **ddom** in the Command window
causes the statements in the script file **ddom.m** to be
executed. Variables in a script file are global and will
change the value of variables of the same name in the
workspace of the current MATLAB session. Type:

```
A = rand (3)
ddom
A
```

in the Command window. It seems to work; the matrix A
is now diagonally dominant. If you type this in the
Command window, though,

```
A = [1 -2 ; -1 1]
ddom
A
```

then the diagonal of A just got worse. What happened?
Click on the Editor window and move the mouse to point
to the variable f, anywhere in the script. You will see a
yellow pop-up window with:

---

[3] See http://www.cise.ufl.edu/research/sparse/MATLAB for the
M-files and MEX-files used in this book.

```
f =
      -1
       0
```

Oops. f is supposed to be a sum of absolute values, so it cannot be negative. Edit the first line of **ddom.m** and change it to:

```
f = sum (abs (A), 2) ;
```

save the file, and run it again on the original matrix A=[1 −2;−1 1]. This time, instead of typing in the command, try running the script by clicking:

in the Editor window. This is a shortcut to typing ddom in the Command window. The matrix A is now diagonally dominant. Run the script again, though, and you will see that A is modified even if it is already diagonally dominant. Fix this modifying only those rows that violate diagonal dominance.

Set A to [1 −2;−1 1] by clicking on the command in the Command History window. Next, modify **ddom.m** to be:

```
d = diag (A) ;
a = abs (d) ;
f = sum (abs (A), 2) - a ;
i = find (f >= a) ;
A (i,i) = A (i,i) + diag (f (i)) ;
```

and click:

33

to save and run the script. Run it again; the matrix does not change.

Try it on the matrix A=[-1 2;1 -1]. The result is wrong. To fix it, try another debugging method — setting breakpoints. A breakpoint causes the script to pause, and allows you to enter commands in the Command window, while the script is paused (it acts just like the keyboard command).

Click on line 5 and select Breakpoints ▶ Set/Clear Breakpoint or click:

A red dot appears in a column to the left of line 5. You can also set and clear breakpoints by clicking on the red dots or dashes in this column.

In the Command window, type:

```
clear
A = [-1 2 ; 1 -1]
ddom
```

A green arrow appears at line 5, and the prompt K>> appears in the Command window. Execution of the script has paused, just before line 5 is executed. Look at the variables A and f. Since the diagonal is negative, and f is an absolute value, we should subtract f from A to preserve the sign. Type the command:

```
A = A - diag (f)
```

The matrix is now correct, although this works only if all of the rows need to be fixed and all diagonal entries are negative. Stop the script by selecting Debug ▸ Exit Debug Mode or by clicking:

Clear the breakpoint. Edit the script, and replace line 5 with:

```
s = sign (d (i)) ;
A (i,i) = A (i,i) + diag (s .* f (i)) ;
```

Type A=[-1 2;1 -1] and run the script. The script seems to work, but it modifies A more than is needed. Try the script on A=zeros(4), and you will see that the matrix is not modified at all, because sign(0) is zero. Fix the script so that it looks like this:

```
d = diag (A) ;
a = abs (d) ;
f = sum (abs (A), 2) - a ;
i = find (f >= a) ;
[m n] = size (A) ;
k = i + (i-1)*m ;
tol = 100 * eps ;
s = 2 * (d (i) >= 0) - 1 ;
A (k) = (1+tol) * s .* max (f (i), tol);
```

which is the sequence of commands you typed in Section 5.5.

7.3 Function files

Function files provide extensibility to MATLAB. You can create new functions specific to your problem, which will then have the same status as other MATLAB functions. Variables in a function file are by default local. A variable can, however, be declared global (see `help global`).

Convert your `ddom.m` script into a function by adding these lines at the beginning of `ddom.m`:

```
function B = ddom (A)
% B = ddom (A) returns a diagonally
% dominant matrix B by modifying the
% diagonal of A.
```

and add this line at the end of your new function:

```
B = A ;
```

You now have a MATLAB function, with one input argument and one output argument. To see the difference between global and local variables as you do this exercise, type `clear`. Functions do not modify their inputs, so:

```
C = [1 -2 ; -1 1]
D = ddom (C)
```

returns a matrix C that is diagonally dominant. The matrix C in the workspace does not change, although a copy of it local to the `ddom` function, called A, is modified as the function executes. Note that the other variables, a, d, f, i, k and s no longer appear in your workspace. Neither do A and B. These are all local to the `ddom` function.

The first line of the function declares the function name, input arguments, and output arguments; without this line the file would be a script file. Then a MATLAB statement D=ddom(C), for example, causes the matrix C to be passed as the variable A in the function and causes the output result to be passed out to the variable D. Since variables in a function file are local, their names are independent of those in the current MATLAB workspace. Your workspace will have only the matrices C and D. If you want to modify C itself, then use C=ddom(C).

Lines that start with % are comments; more on this in Section 7.6. An optional return statement causes the function to finish and return its outputs.

7.4 Multiple inputs and outputs

A function may also have multiple output arguments. For example, it would be useful to provide the caller of the ddom function some control over how strong the diagonal is to be and to provide more results, such as the list of rows (the variable i) that violated diagonal dominance. Try changing the first line to:

```
function [B,i] = ddom (A, tol)
```

and add a % at the beginning of the line that computes tol. Single assignments can also be made with a function having multiple output arguments. For example, with this version of ddom, the statement D=ddom(C,0.1) will assign the modified matrix to the variable D without returning the vector i. Try it.

7.5 Variable arguments

Not all inputs and outputs of a function need be present when the function is called. The variables nargin and nargout can be queried to determine the number of inputs and outputs present. For example, we could use a default tolerance if `tol` is not present. Add these statements in place of the line that computed `tol`:

```
if (nargin == 1)
    tol = 100 * eps ;
end
```

An example of both `nargin` and `nargout` is given in Section 8.1.

7.6 Comments and documentation

The % symbol indicates that the rest of the line is a comment; MATLAB will ignore the rest of the line. Moreover, the first contiguous comment lines are used to document the M-file. They are available to the online help facility and will be displayed if, for example, `help ddom` is entered. Such documentation should always be included in a function file. Since you've modified the function to add new inputs and outputs, edit your script to describe the variables `i` and `tol`. Be sure to state what the default value of `tol` is. Next, type `help ddom`.

7.7 Entering large matrices

Script files may be used to enter data into a large matrix; in such a file, entry errors can be easily corrected. If, for example, one enters in a file `amatrix.m`:

```
A = [
1 2 3 4
5 6 7 8
] ;
```

then the command `amatrix` causes the assignment given in `amatrix.m` to be carried out. However, it is usually easier to use `load` (see Section 2.7) or the Array Editor (see Section 2.6), rather than a script.

An M-file can reference other M-files, including referencing itself recursively.

8. Advanced M-file features

This section describes advanced M-file techniques, such as how to pass function references and how to write high-performance code in MATLAB.

8.1 Function references

A function handle is a reference to a function that can then be treated as a variable. It can be copied, stored in a matrix (not a numeric one, though), placed in cell array, and so on. Its final use is normally to pass it to `feval`, which then evaluates the function. For example,

```
h = @sin
y = feval (h, pi/2)
```

is the same thing as simply y=sin(pi/2). Try it. You can also use a string to refer to a function, as in:

```
y = feval ('sin', pi/2)
```

but the function handle method is more general. See `help function_handle` for more information.

The `bisect` function, below, takes a function handle as one of its inputs. It also gives you an example of `nargin` and `nargout` (see also Section 7.5).

```
function [b, steps] = bisect (fun,x,tol)
% BISECT:  zero of a function of one
% variable via the bisection method.
% bisect (fun,x) returns a zero of the
% function fun.  fun is a function
% handle or a string with the name of a
```

```
% function.  x is a starting guess. The
% value of b returned is near a point
% where fun changes sign.  For example,
% bisect (@sin,3) is pi.  Note the use
% of the function handle, @sin.
%
% An optional third input argument sets
% a tolerance for the relative accuracy
% of the result.  The default is eps.
% An optional second output argument
% gives a matrix containing a trace of
% the steps; the rows are of the form
% [c (f(c))].

if (nargin < 3)
    % default tolerance
    tol = eps ;
end
trace = (nargout == 2) ;
if (x ~= 0)
    dx = x/20 ;
else
    dx = 1/20 ;
end
a = x - dx ;
fa = feval (fun, a) ;
b = x + dx ;
fb = feval (fun, b) ;
if (trace)
    steps = [a fa ; b fb] ;
end

% find a change of sign
while (fa > 0) == (fb > 0)
    dx = 2*dx ;
    a = x - dx ;
    fa = feval (fun, a) ;
    if (trace)
        steps = [steps ; [a fa]] ;
    end
    if (fa > 0) ~= (fb > 0)
        break
    end
```

40

```
        b = x + dx ;
        fb = feval (fun, b) ;
        if (trace)
            steps = [steps ; [b fb]] ;
        end
    end

    % main loop
    while (abs (b-a) > 2*tol*max(abs(b),1))
        c = a + (b-a)/2 ;
        fc = feval (fun, c) ;
        if (trace)
            steps = [steps ; [c fc]] ;
        end
        if (fb > 0) == (fc > 0)
            b = c ;
            fb = fc ;
        else
            a = c ;
            fa = fc ;
        end
    end
```

Some of MATLAB's functions are built in; others are
distributed as M-files. The actual listing of any
non-built-in M-file, MATLAB's or your own, can be
viewed with the MATLAB command type
functionname. Try entering type eig, type vander,
and type rank.

8.2 Name resolution

When MATLAB comes upon a new name, it resolves it
into a specific variable or function by checking to see if it
is a variable, a built-in function, a file in the current
directory, or a file in the MATLAB path (in order of the
directories listed in the path). MATLAB uses the first
variable, function, or file it encounters with the specified
name. There are other cases; see Help: MATLAB: Using

MATLAB: Development Environment: Workspace, Path, and File Operations: Search Path. You can use the command which to find out what a name is. Try this:

```
clear
i
which i
i = 3
which i
```

8.3 Error messages

Error messages are best displayed with the function error. For example,

```
A = rand (4,3)
[m n] = size (A) ;
if m ~= n
    error ('A must be square') ;
end
```

aborts execution of an M-file if the matrix A is not square. This is a useful thing to add to the ddom function that you developed in Chapter 7, since diagonal dominance is only defined for square matrices. Try adding it to ddom (excluding the rand statement, of course), and see what happens if you call ddom with a rectangular matrix.

See Section 6.5 (try/catch) for one way to deal with errors in functions you call.

8.4 User input

In an M-file the user can be prompted to interactively enter input data, expressions, or commands. When, for example, the statement:

```
iter = input ('iteration count: ') ;
```

is encountered, the prompt message is displayed and execution pauses while the user keys in the input data (or, in general, any MATLAB expression). Upon pressing the return key, the data is assigned to the variable iter and execution resumes. You can also input a string; see help input.

An M-file can be paused until a return is typed in the Command window with the pause command. It is a good idea to display a message, as in:

```
disp ('Hit enter to continue: ') ;
pause
```

A Ctrl-C will terminate the script or function that is paused. A more general command, keyboard, allows you to type any number of MATLAB commands. See help keyboard.

8.5 Efficient code

The function ddom.m that you wrote in Chapter 7 illustrates some of the MATLAB features that can be used to produce efficient code. All operations are "vectorized," and loops are avoided. We could have written the ddom function using nested for loops, much like how you would write it in C, FORTRAN, or Java:

```
function B = ddom (A,tol)
% B = ddom (A) returns a diagonally
% dominant matrix B by modifying the
% diagonal of A.
[m n] = size (A) ;
if (nargin == 1)
    tol = 100 * eps ;
end
for i = 1:n
    d = A (i,i) ;
```

```
                a = abs (d) ;
                f = 0 ;
                for j = 1:n
                    if (i ~= j)
                        f = f + abs (A (i,j)) ;
                    end
                end
                if (f >= a)
                    aii = (1 + tol) * max (f, tol) ;
                    if (d < 0)
                        aii = -aii ;
                    end
                    A (i,i) = aii ;
                end
            end
        B = A ;
```

This works, but it is very slow for large matrices. As you become practiced in writing without loops and reading loop-free MATLAB code, you will also find that the loop-free version is easier to read and understand.

If you cannot vectorize some computations, you can make your for loops go faster by preallocating any vectors or matrices in which output is stored. For example, by including the second statement below, which uses the function zeros, space for storing E in memory is preallocated. Without this, MATLAB must resize E one column larger in each iteration, slowing execution.

```
M = magic (6) ;
E = zeros (6,50) ;
for j = 1:50
    E (:,j) = eig (M^j) ;
end
```

8.6 Performance measures

Time and space are the two basic measures of an algorithm's efficiency. In MATLAB, this translates into

44

the number of floating-point operations (flops) performed, the elapsed time, the CPU time, and the memory space used. MATLAB no longer provides a flop count because it uses high-performance block matrix algorithms that make it difficult to count the actual flops performed. See `help flops`.

The elapsed time (in seconds) can be obtained with the stopwatch timers `tic` and `toc`; `tic` starts the timer and `toc` returns the elapsed time. Hence, the commands:

```
tic
statement
toc
```

will return the elapsed time for execution of the *statement*. The elapsed time for solving a linear system above can be obtained, for example, with:

```
n = 500 ;
A = rand (n) ;
b = rand (n,1) ;
tic
x = A\b ;
toc
r = norm (A*x-b)
```

The norm of the residual is also computed. You may wish to compare x=A\B with x=inv(A)*b for solving the linear system. Try it. You will generally find A\b to be faster and more accurate.

If there are other programs running at the same time on your computer, elapsed time will not be an accurate measure of performance. Try using `cputime` instead. See `help cputime`.

MATLAB runs faster if you can restructure your computations to use less memory. Type the following and select n to be some large integer, such as:

```
n = 16000 ;
a = rand (n,1) ;
b = rand (1,n) ;
c = rand (n,1) ;
```

Here are three ways of computing the same vector x. The first one uses hardly any extra memory, the second and third use a huge amount (about 2GB). Try them (good luck!).

```
x = a*(b*c) ;
x = (a*b)*c ;
x = a*b*c ;
```

No measure of peak memory usage is provided. You can find out the total size of your workspace, in bytes, with the command whos. The total can also be computed with:

```
s = whos
space = sum ([s.bytes])
```

Try it. This does not give the peak memory used while inside a MATLAB operator or function, though. See help memory for more options.

8.7 Profile

MATLAB provides an M-file profiler that lets you see how much computation time each line of an M-file uses. The command to use is profile (see help profile for details).

9. Calling C from MATLAB

There are times when MATLAB itself is not enough.
You may have a large application or library written in
another language that you would like to use from
MATLAB, or it might be that the performance of your M-
file is not what you would like.

MATLAB can call routines written in C, FORTRAN, or
Java. Similarly, programs written in C and FORTRAN
can call MATLAB. In this chapter, we will just look at
how to call a C routine from MATLAB. For more
information, see Help: MATLAB: External
Interfaces/API, or see the online MATLAB
document *External Interfaces*. This discussion assumes
that you already know C.

9.1 A simple example

A routine written in C that can be called from MATLAB
is called a MEX-file. The routine must always have the
name mexFunction, and the arguments to this routine
are always the same. Here is a very simple MEX-file;
type it in as the file hello.c in your favorite text editor.

```
#include "mex.h"
void mexFunction
(
    int nlhs,
    mxArray *plhs [ ],
    int nrhs,
    const mxArray *prhs [ ]
)
{
    mexPrintf ("hello world\n") ;
}
```

Compile and run it by typing:

```
mex hello.c
hello
```

If this is the first time you have compiled a C MEX-file on a PC with Microsoft Windows, you will be prompted to select a C compiler. MATLAB for the PC comes with its own C compiler (lcc). The arguments nlhs and nrhs are the number of outputs and inputs to the function, and plhs and prhs are pointers to the arguments themselves (of type mxArray). This hello.c MEX-file does not have any inputs or outputs, though.

The mexPrintf function is just the same as printf. You can also use printf itself; the mex command redefines it as mexPrintf when the program is compiled. This way, you can write a routine that can be used from MATLAB or from a stand-alone C application, without MATLAB.

9.2 C versus MATLAB arrays

MATLAB stores its arrays in column major order, while the convention for C is to store them in row major order. Also, the number of columns in an array is not known until the mexFunction is called. Thus, two-dimensional arrays in MATLAB must be accessed with one-dimensional indexing in C (see also Section 5.5). In the example in the next section, the INDEX macro helps with this translation.

Array indices also appear differently. MATLAB is written in C, and it stores all of its arrays internally using zero-based indexing. An m-by-n matrix has rows 0 to m-1 and columns 0 to n-1. However, the user interface to these arrays is always one-based, and index vectors in

MATLAB are always one-based. In the example below, one is added to the List array returned by diagdom to account for this difference.

9.3 A matrix computation in C

In Chapters 7 and 8, you wrote the function ddom.m. Here is the same function written as an ANSI C MEX-file. Compare the diagdom routine, below, with the loop-based version of ddom.m in Section 8.5. The MATLAB mx and mex routines are described in Section 9.4. To save space, the comments are terse.

```c
#include "mex.h"
#include "matrix.h"
#include <stdlib.h>
#include <float.h>
#define INDEX(i,j,m) ((i)+(j)*(m))
#define ABS(x) ((x) >= 0 ? (x) : -(x))
#define MAX(x,y) (((x)>(y)) ? (x):(y))

void diagdom
(
    double *A,
    int n,
    double *B,
    double tol,
    int *List,
    int *nList
)
{
    int i, j, k ;
    double d, a, f, bij, bii ;
    for (k = 0 ; k < n*n ; k++)
    {
        B [k] = A [k] ;
    }
    if (tol < 0)
    {
        tol = 100 * DBL_EPSILON ;
    }
```

49

```
    k = 0 ;
    for (i = 0 ; i < n ; i++)
    {
        d = B [INDEX (i,i,n)] ;
        a = ABS (d) ;
        f = 0 ;
        for (j = 0 ; j < n ; j++)
        {
            if (i != j)
            {
                bij = B [INDEX (i,j,n)];
                f += ABS (bij) ;
            }
        }
        if (f >= a)
        {
            List [k++] = i ;
            bii = (1 + tol) *
                    MAX (f, tol) ;
            if (d < 0)
            {
                bii = -bii ;
            }
            B [INDEX (i,i,n)] = bii ;
        }
    }
    *nList = k ;
}

void error (char *s)
{
    mexPrintf ("Usage: [B,i] = "
    "diagdom (A,tol)\n") ;
    mexErrMsgTxt (s) ;
}

void mexFunction
(
    int nlhs,
    mxArray *plhs [ ],
    int nrhs,
    const mxArray *prhs [ ]
)
```

50

```
{
    int n, k, *List, nList ;
    double *A, *B, *I, tol ;

    /* get inputs A and tol */
    if (nlhs > 2 || nrhs > 2
    || nrhs == 0)
    {
        error (
        "Wrong number of arguments") ;
    }
    if (mxIsEmpty (prhs [0]))
    {
        plhs [0] = mxCreateDoubleMatrix
                   (0, 0, mxREAL) ;
        plhs [1] = mxCreateDoubleMatrix
                   (0, 0, mxREAL) ;
        return ;
    }
    n = mxGetN (prhs [0]) ;
    if (n != mxGetM (prhs [0]))
    {
        error ("A must be square") ;
    }
    if (mxIsSparse (prhs [0]))
    {
        error ("A cannot be sparse") ;
    }
    A = mxGetPr (prhs [0]) ;
    tol = -1 ;
    if (nrhs > 1
    && !mxIsEmpty (prhs [1]))
    {
        tol = mxGetScalar (prhs [1]) ;
    }

    /* create output B */
    plhs [0] = mxCreateDoubleMatrix
               (n, n, mxREAL) ;
    B = mxGetPr (plhs [0]) ;

    /* get temporary workspace */
    List = (int *) mxMalloc
```

51

```
            (n * sizeof (int)) ;

    /* do the computation */
    diagdom (A, n, B,tol, List, &nList);

    /* create output I */
    plhs [1] = mxCreateDoubleMatrix
                (nList, 1, mxREAL) ;
    I = mxGetPr (plhs [1]) ;
    for (k = 0 ; k < nList ; k++)
    {
        I [k] = (double) (List[k] + 1);
    }

    /* free the workspace */
    mxFree (List) ;
}
```

Type it in as the file diagdom.c (or get it from the web),
and then type:

```
mex diagdom.c
A = rand (6) ;
B = ddom (A) ;
C = diagdom (A) ;
```

The matrices B and C will be the same (round-off error
might cause them to differ slightly).

9.4 MATLAB mx and mex routines

In the last example, the C routine calls several routines
with the prefix mx or mex. These are routines in
MATLAB. Routines with mx prefixes operate on
MATLAB matrices and include:

mxIsEmpty	1 if the matrix is empty, 0 otherwise
mxIsSparse	1 if the matrix is sparse, 0 otherwise
mxGetN	number of columns of a matrix
mxGetM	number of rows of a matrix

mxGetPr	pointer to the real values of a matrix
mxGetScalar	the value of a scalar
mxCreateDoubleMatrix	create MATLAB matrix
mxMalloc	like malloc in ANSI C
mxFree	like free in ANSI C

Routines with mex prefixes operate on the MATLAB environment and include:

mexPrintf	like printf in C
mexErrMsgTxt	like MATLAB's error statement
mexFunction	the gateway routine from MATLAB

You will note that all of the references to MATLAB's mx and mex routines are limited to the mexFunction gateway routine. This is not required; it is just a good idea. Many other mx and mex routines are available.

The memory management routines in MATLAB (mxMalloc, mxFree, and mxCalloc) are much easier to use than their ANSI C counterparts. If a memory allocation request fails, the mexFunction terminates and control is passed backed to MATLAB. Any workspace allocated by mxMalloc that is not freed when the mexFunction returns or terminates is automatically freed by MATLAB. This is why no memory allocation error checking is included in diagdom.c; it is not necessary.

9.5 Online help for MEX routines

Create an M-file called diagdom.m, with only this:

```
function [B,i] = diagdom (A,tol)
% diagom:  modify the matrix A
% [B,i] = diagdom (A,tol) returns a
```

53

```
% diagonally dominant matrix B by
% modifying the diagonal of A.
error ('diagdom mexFunction not found');
```

Now type `help diagdom`. This is a simple method for
providing online help for your own MEX-files.

9.6 Larger examples on the web

The `colamd` and `symamd` routines in MATLAB are C
MEX-files. The source code for these routines is on the
web at http://www.cise.ufl.edu/research/sparse/colamd.
Like the example in the previous section, they are split
into a `mexFunction` gateway routine and another set of
routines that do not make use of MATLAB.

10. Two-Dimensional Graphics

MATLAB can produce two-dimensional plots. The
primary command for this is `plot`. Chapter 11 discusses
three-dimensional graphics. To preview some of these
capabilities, enter the command `demo` and select some of
the visualization and graphics demos.

10.1 Planar plots

The `plot` command creates linear x–y plots; if x and y
are vectors of the same length, the command `plot(x,y)`
opens a graphics window and draws an x–y plot of the
elements of y versus the elements of x. You can, for
example, draw the graph of the sine function over the
interval −4 to 4 with the following commands:

```
x = -4:0.01:4 ;
y = sin (x) ;
plot (x, y) ;
```

Try it. The vector x is a partition of the domain with mesh size 0.01, and y is a vector giving the values of sine at the nodes of this partition (recall that sin operates entry-wise). When plotting a curve, the plot routine is actually connecting consecutive points induced by the partition with line segments. Thus, the mesh size should be chosen sufficiently small to render the appearance of a smooth curve.

You will usually want to keep the current Figure window exposed, but moved to the side, and the Command window active.

As a second example, draw the graph of $y = e^{-x^2}$ over the interval -1.5 to 1.5 as follows:

```
x = -1.5:.01:1.5 ;
y = exp (-x.^2) ;
plot (x, y) ;
```

Note that you must precede ^ by a period to ensure that it operates entry-wise.

Select Tools ▶ Zoom In or Tools ▶ Zoom Out in the Figure window to zoom in or out of the plot. See also the zoom command (help zoom).

10.2 Multiple figures

You can have several concurrent Figure windows, one of which will at any time be the designated current figure in which graphs from subsequent plotting commands will be placed. If, for example, Figure 1 is the current figure, then the command figure(2) (or simply figure) will open a second figure (if necessary) and make it the current figure. The command figure(1) will then

55

expose Figure 1 and make it again the current figure. The command gcf returns the current figure number.

MATLAB does not draw a plot right away. It waits until all computations are finished, until a figure command is encountered, or until the script or function requests user input (see Section 8.4). To force MATLAB to draw a plot right away, use the command figure(gcf). This does not change the current figure.

10.3 Graph of a function

MATLAB supplies a function fplot to easily and efficiently plot the graph of a function. For example, to plot the graph of the function above, you can first define the function in an M-file called, say, expnormal.m containing:

```
function y = expnormal (x)
y = exp(-x.^2) ;
```

Then either of the commands:

```
fplot ('expnormal', [-1.5 1.5]) ;
fplot (@expnormal, [-1.5 1.5]) ;
```

will produce the graph over the indicated x-domain. The first one uses a string to refer to the function. The second one uses a function handle (which is preferred). Try it.

A faster way to see the same result without creating expnormal.m would be:

```
fplot ('exp(-x^2)', [-1.5 1.5]) ;
```

The variable x in the expression above is a place-holder; it need not exist and can be any arbitrary variable name.

10.4 Parametrically defined curves

Plots of parametrically defined curves can also be made. Try, for example,

```
t = 0:.001:2*pi ;
x = cos (3*t) ;
y = sin (2*t) ;
plot (x, y) ;
```

10.5 Titles, labels, text in a graph

The graphs can be given titles, axes labeled, and text placed within the graph with the following commands, which take a string as an argument.

title	graph title
xlabel	x-axis label
ylabel	y-axis label
gtext	place text on graph using the mouse
text	position text at specified coordinates

For example, the command:

```
title ('A parametric cos/sin curve')
```

gives a graph a title. The command gtext('The Spot') lets you interactively place the designated text on the current graph by placing the mouse crosshair at the desired position and clicking the mouse. It is a good idea to prompt the user before using gtext. To place text in a graph at designated coordinates, use the command text (see help text). These commands are also in the Insert menu in the Figure window. Select Insert ▶ Text, click on the figure, type something, and then click somewhere else to finish entering the text. If the edit-figure button:

is depressed (or select Tools ▸ Edit Plot), you can
right-click on anything in the figure and see a pop-up
menu that gives you options to modify the item you just
clicked. You can also click and drag objects on the
figure. Selecting Edit ▸ Axes Properties brings up a
window with many more options. For example, clicking
the:

Grid ☑ Show

box adds grid lines (the command grid does the same
thing).

10.6 Control of axes and scaling

By default, the axes are auto-scaled. This can be
overridden by the command axis or by selecting Edit ▸
Axes Properties. Some features of the axis
command are:

```
axis ([xmin xmax ymin ymax])
                      sets the axes
axis manual           freezes the current axes for
                      new plots
axis auto             returns to auto-scaling
v = axis              vector v shows current scaling
axis square           axes same size (but not scale)
axis equal            same scale and tic marks on axes
axis off              removes the axes
axis on               restores the axes
```

The `axis` command should be given after the `plot`
command. Try `axis([-2 2 -3 3])` with the current
figure. You will note that text entered on the figure using
the `text` or `gtext` moves as the scaling changes (think
of it as attached to the data you plotted). Text entered via
`Insert ▶ Text` stays put.

10.7 Multiple plots

Two ways to make multiple plots on a single graph are
illustrated by:

```
x = 0:.01:2*pi;
y1 = sin (x) ;
y2 = sin (2*x) ;
y3 = sin (4*x) ;
plot (x, y1, x, y2, x, y3)
```

and by forming a matrix `Y` containing the functional
values as columns:

```
x = 0:.01:2*pi ;
Y = [sin(x)', sin(2*x)', sin(4*x)'] ;
plot (x, Y)
```

The x and y pairs must have the same length, but each
pair can have different lengths. Try:

```
plot (x, Y, [0 2*pi], [0 0])
```

The command `hold on` freezes the current graphics
screen so that subsequent plots are superimposed on it.
The axes may, however, become rescaled. Entering `hold
off` releases the hold.

The function `legend` places a legend in the current figure
to identify the different graphs. See `help legend`.

Clearing a figure can be done with clf, which clears the axes, the data you plotted, any text entered with the text and gtext commands, and the legend. To also clear the text you entered via Insert ▶ Text, type clf reset.

10.8 Line types, marker types, colors

You can override the default line types, marker types, and colors. For example,

```
x = 0:.01:2*pi ;
y1 = sin (x) ;
y2 = sin (2*x) ;
y3 = sin (4*x) ;
plot (x,y1, '--', x,y2, ':', x,y3, '+')
```

renders a dashed line and dotted line for the first two graphs, whereas for the third the symbol + is placed at each node. The line types are:

'-'	solid	':'	dotted
'--'	dashed	'-.'	dashdot

and the marker types are:

'.'	point	'o'	circle
'x'	x-mark	'+'	plus
'*'	star	's'	square
'd'	diamond	'v'	triangle-down
'^'	triangle-up	'<'	triangle-left
'>'	triangle-right	'p'	pentagram
'h'	hexagram		

Colors can be specified for the line and marker types:

'y'	yellow	'm'	magenta
'c'	cyan	'r'	red

| 'g' | green | 'b' | blue |
| 'w' | white | 'k' | black |

For example, `plot(x,y1,'r--')` plots a red dashed line.

10.9 Subplots and specialized plots

The command `subplot` partitions a figure so that several small plots can be placed in one figure. See `help subplot`. Other specialized planar plotting functions you may wish to explore via `help` are:

bar	fill	quiver
compass	hist	rose
feather	polar	stairs

10.10 Graphics hard copy

Select File ▶ Print or click the print button:

in the Figure window to send a copy of your figure to your default printer. Layout options and selecting a printer can be done with File ▶ Page Setup and File ▶ Print Setup.

You can save the figure as a file for later use in a MATLAB Figure window. Try the save button:

or File ▶ Save. This saves the figure as a `.fig` file, which can be later opened in the Figure window with the open button:

or with File ▶ Open. Selecting File ▶ Export allows
you to convert your figure to many other formats.

11. Three-Dimensional Graphics

MATLAB's primary commands for creating three-
dimensional graphics are plot3, mesh, surf, and
light. The menu options and commands for setting
axes, scaling, and placing text, labels, and legends on a
graph also apply for three-dimensional graphs. A
zlabel can be added. The axis command requires a
vector of length 6 with a 3-D graph.

11.1 Curve plots

Completely analogous to plot in two dimensions, the
command plot3 produces curves in three-dimensional
space. If x, y, and z are three vectors of the same size,
then the command plot3(x,y,z) produces a
perspective plot of the piecewise linear curve in
three-space passing through the points whose coordinates
are the respective elements of x, y, and z. These vectors
are usually defined parametrically. For example,

```
t = .01:.01:20*pi ;
x = cos (t) ;
y = sin (t) ;
z = t.^3 ;
plot3 (x, y, z)
```

produces a helix that is compressed near the x–y plane (a
"slinky"). Try it.

11.2 Mesh and surface plots

The mesh command draws three-dimensional wire mesh surface plots. The command mesh(z) creates a three-dimensional perspective plot of the elements of the matrix z. The mesh surface is defined by the z-coordinates of points above a rectangular grid in the x–y plane. Try mesh(eye(20)).

Similarly, three-dimensional faceted surface plots are drawn with the command surf. Try surf(eye(20)).

To draw the graph of a function $z = f(x, y)$ over a rectangle, first define vectors xx and yy, which give partitions of the sides of the rectangle. The function meshgrid(xx,yy) then creates a matrix x, each row of which equals xx (whose column length is the length of yy) and similarly a matrix y, each column of which equals yy. A matrix z, to which mesh or surf can be applied, is then computed by evaluating the function f entry-wise over the matrices x and y.

You can, for example, draw the graph of $z = e^{-x^2 - y^2}$ over the square [-2, 2] x [-2, 2] as follows (try it):

```
xx = -2:.2:2 ;
yy = xx ;
[x, y] = meshgrid (xx, yy) ;
z = exp (-x.^2 - y.^2) ;
mesh (z)
```

Try this plot with surf instead of mesh. Note that you must use x.^2 and y.^2 instead of x^2 and y^2 to ensure that the function acts entry-wise on x and y.

11.3 Color shading and color profile

The color shading of surfaces is set by the `shading` command. There are three settings for shading: `faceted` (default), `interpolated`, and `flat`. These are set by the commands:

```
shading faceted
shading interp
shading flat
```

Note that on surfaces produced by `surf`, the settings `interpolated` and `flat` remove the superimposed mesh lines. Experiment with various shadings on the surface produced above. The command `shading` (as well as `colormap` and `view` described below) should be entered after the `surf` command.

The color profile of a surface is controlled by the `colormap` command. Available predefined color maps include `hsv` (the default), `hot`, `cool`, `jet`, `pink`, `copper`, `flag`, `gray`, `bone`, `prism`, and `white`. The command `colormap(cool)`, for example, sets a certain color profile for the current figure. Experiment with various color maps on the surface produced above. See also `help colorbar`.

11.4 Perspective of view

The Figure window provides a wide range of controls for viewing the figure. Select `View ▶ Camera Toolbar` to see these controls, or pull down the `Tools` menu. Try, for example, selecting `Tools ▶ Rotate 3-D`, and then click the mouse in the Figure window and drag it to rotate the object. Some of these options can be controlled by the `view` and `rotate3d` commands, respectively.

The MATLAB function peaks generates an interesting surface on which to experiment with shading, colormap, and view. Type peaks, select Tools ▶ Rotate 3-D, and click and drag the figure to rotate it.

In MATLAB, light sources and camera position can be set. Taking the peaks surface from the example above, select Insert ▶ Light, or type light to add a light source. See the online document *Using MATLAB Graphics* for camera and lighting help.

11.5 Parametrically defined surfaces

Plots of parametrically defined surfaces can also be made. The MATLAB functions sphere and cylinder generate such plots of the named surfaces. (See type sphere and type cylinder.) The following is an example of a similar function that generates a plot of a torus by utilizing spherical coordinates.

```
function [x, y, z] = torus (r, n, a)
% TORUS Generate a torus.
% torus (r, n, a) generates a plot of a
% torus with central radius a and
% lateral radius r.  n controls the
% number of facets on the surface.
% These input variables are optional
% with defaults r = 0.5, n = 30, a = 1.
% [x, y, z] = torus(r, n, a) generates
% three (n + 1)-by-(n + 1) matrices so
% that surf (x, y, z) will produce the
% torus.  See also SPHERE, CYLINDER.
if nargin < 3, a = 1 ; end
if nargin < 2, n = 30 ; end
if nargin < 1, r = 0.5 ; end
theta = pi * (0:2:2*n)/n ;
phi = 2*pi* (0:2:n)'/n ;
xx = (a + r*cos(phi)) * cos(theta) ;
yy = (a + r*cos(phi)) * sin(theta) ;
```

```
zz = r * sin(phi) * ones(size(theta)) ;
if nargout == 0
    surf (xx, yy, zz) ;
    ar = (a + r)/sqrt(2) ;
    axis([-ar, ar, -ar, ar, -ar, ar]) ;
else
    x = xx ;
    y = yy ;
    z = zz ;
end
```

Other three-dimensional plotting functions you may wish
to explore via help are meshz, surfc, surfl, contour,
and pcolor.

12. Advanced Graphics

MATLAB possesses a number of other advanced
graphics capabilities. Significant ones are object-based
graphics, called Handle Graphics, and Graphical User
Interface (GUI) tools.

12.1 Handle Graphics

Beyond those just described, MATLAB's graphics
system provides low-level functions that let you control
virtually all aspects of the graphics environment to
produce sophisticated plots. The commands set and get
allow access to all the properties of your plots. Try
set(gcf) to see some of the properties of a figure that
you can control. This system is called Handle Graphics.
See *Using MATLAB Graphics* for more information.

12.2 Graphical user interface

MATLAB's graphics system also provides the ability to
add sliders, push-buttons, menus, and other user interface
controls to your own figures. For information on creating
user interface controls, try help uicontrol. This

allows you to create interactive graphical-based applications.

Try `guide` (short for Graphic User Interface Development Environment). This brings up MATLAB's Layout Editor window that you can use to interactively design a graphic user interface.

For more information, see the online document *Creating Graphical User Interfaces*.

13. Sparse Matrix Computations

A sparse matrix is one with mostly zero entries. MATLAB provides the capability to take advantage of the sparsity of matrices.

13.1 Storage modes

MATLAB has two storage modes, full and sparse, with full the default. The functions `full` and `sparse` convert between the two modes. Nearly all MATLAB operators and functions operate seamlessly on both full and sparse matrices. For a matrix A, full or sparse, `nnz(A)` returns the number of nonzero elements in A.

An m-by-n sparse matrix is stored in three one-dimensional arrays. Numerical values and their row indices are stored in two arrays of size `nnz(A)` each. All of the entries in any given column are stored contiguously. A third array of size n+1 holds the positions in the other two arrays of the first nonzero entry in each column. Thus, if A is sparse, then x=A(9,:) takes much more time than x=A(:,9), and s=A(4,5) is also slow. To get high performance when dealing with sparse matrices, use matrix expressions instead of `for` loops and vector or scalar expressions. If you must operate on the rows of a sparse matrix A, try working with the columns of A' instead.

If a full tridiagonal matrix F is created via, say,

```
F = floor (10 * rand(6)) ;
F = triu (tril (F,1), -1) ;
```

then the statement S=sparse(F) will convert F to sparse mode. Try it. Note that the output lists the nonzero entries in column major order along with their row and column indices because of how sparse matrices are stored. The statement F=full(S) returns F in full storage mode. You can check the storage mode of a matrix A with the command issparse(A).

13.2 Generating sparse matrices

A sparse matrix is usually generated directly rather than by applying the function sparse to a full matrix. A sparse banded matrix can be easily created via the function spdiags by specifying diagonals. For example, a familiar sparse tridiagonal matrix is created by:

```
m = 6 ;
n = 6 ;
e = ones (n,1) ;
d = -2*e ;
T = spdiags ([e d e], [-1 0 1], m, n)
```

Try it. The integral vector [-1 0 1] specifies in which diagonals the columns of [e d e] should be placed (use full(T) to see the full matrix T and spy(T) to view T graphically). Experiment with other values of m and n and, say, [-3 0 2] instead of [-1 0 1]. See help spdiags for further features of spdiags.

The sparse analogs of `eye`, `zeros`, `ones`, and `rand` for full matrices are, respectively, `speye`, `sparse`, `spones`, and `sprand`. The latter two take a matrix argument and replace only the nonzero entries with ones and uniformly distributed random numbers, respectively. `sparse(m,n)` creates a sparse zero matrix. `sprand` also permits the sparsity structure to be randomized. This is a useful method for generating simple sparse test matrices, but be careful. Random sparse matrices are not truly "sparse" because of catastrophic fill-in when they are factorized (see Section 13.4). Sparse matrices arising in real applications typically do not share this characteristic.[4]

The versatile function `sparse` also permits creation of a sparse matrix via listing its nonzero entries:

```
i = [1 2 3 4 4 4] ;
j = [1 2 3 1 2 3] ;
s = [5 6 7 8 9 10] ;
S = sparse (i, j, s, 4, 3)
full (S)
```

The last two arguments to `sparse` in the example above are optional. They tell `sparse` the dimensions of the matrix; if not present, then S will be max(i) by max(j). If there are repeated entries in [i j], then the entries are added together. The commands below create a matrix whose diagonal entries are 2, 1, and 1.

```
i = [1 2 3 1] ;
j = [1 2 3 1] ;
s = [1 1 1 1] ;
S = sparse (i, j, s)
full (S)
```

---

[4] See http://www.cise.ufl.edu/research/sparse/matrices for a wide range of sparse matrices arising in real applications.

The entries in i, j, and s can be in any order (the same order for all three arrays, of course). In general, if the vector s lists the nonzero entries of S and the integral vectors i and j list their corresponding row and column indices, then:

```
sparse (i, j, s, m, n)
```

will create the desired sparse m-by-n matrix S. As another example try:

```
n = 6 ;
e = floor (10 * rand (n-1,1)) ;
E = sparse (2:n, 1:n-1, e, n, n)
```

13.3 Computation with sparse matrices

The arithmetic operations and most MATLAB functions can be applied independent of storage mode. The storage mode of the result depends on the storage mode of the operands or input arguments. Operations on full matrices always give full results. If F is a full matrix, S and s are sparse, and n is a scalar, then these operations give sparse results:

```
S+S        S*S        S.*S       S.*F
S^n        S.^n       S\s
inv(S)     chol(S)    lu(S)
diag(S)    max(S)     sum(S)
```

These give full results:

```
S+F        F\S        S/F
S*F        S\F        F/S
```

unless F is a scalar, in which case S*F, F\S, and S/F are sparse.

A matrix built from blocks, such as [A, B; C, D], is
stored in sparse mode if any constituent block is sparse.
To compute the eigenvalues or singular values of a sparse
matrix S, you must convert S to a full matrix and then use
eig or svd, as eig(full(S)) or svd(full(S)). If S
is a large sparse matrix and you wish only to compute
some of the eigenvalues or singular values, then you can
use the eigs or svds functions (eigs(S) or svds(S)).

13.4 Ordering methods

When MATLAB solves a sparse linear system (x=A\b), it
typically starts by computing the LU, QR, or Cholesky
factorization of A. This usually leads to fill-in, or the
creation of new nonzeros in the factors that do not appear
in A. MATLAB provides several methods that attempt to
reduce fill-in by reordering the rows and columns of A:

colamd	approximate minimum degree
colmmd	multiple minimum degree
colperm	sort columns by number of nonzeros
symamd	symmetric approximate min. degree
symmmd	symmetric multiple minimum degree
symrcm	reverse Cuthill-McKee

The first three find a column ordering of A and are best
used for lu or qr. The next three are primarily for chol
and return an ordering to be applied symmetrically to
both the rows and columns of a symmetric matrix A (they
can also be used for unsymmetric matrices). Finding the
best ordering is so difficult that it is practically impossible
for most matrices. Fast non-optimal heuristics are used
instead, which means that no one method is always the
best. MATLAB uses colmmd and symmmd by default in

x=A\b, although `colamd` and `symamd` tend to be faster and find better orderings.

Create the `try_lu` function, which also illustrates the use of permutation vectors, the `spy`, `subplot`, `normest`, and `etreeplot` functions, and how to get a close estimate of the flop count for LU factorization if we assume that all zeros are taken advantage of:

```
function try_lu (A, method, issym)
% sparse LU factorization of A
figure (1)
clf reset
subplot (2, 2, 1)
spy (A)
title ('Original matrix A')
t = cputime ;
if (nargin > 2)
    S = spones (A) + spones (A') ;
    p = feval (method, S) ;
    A = A (p,p) ;
elseif (nargin > 1)
    q = feval (method, A) ;
    A = A (:,q) ;
end
torder = cputime - t
subplot (2, 2, 2)
spy (A)
title ('Permuted matrix A')
t = cputime ;
[L, U, P] = lu (A) ;
tlu = cputime - t
total = torder + tlu
subplot (2, 2, 3)
spy (L+U)
title ('LU factors')
normest (L*U-P*A)
Lnz = full (sum (spones (L))) - 1 ;
Unz = full (sum (spones (U')))' - 1 ;
flop_count = 2*Lnz*Unz + sum (Lnz)
subplot (2, 2, 4)
```

```
S = spones (A) ;
etreeplot (S'*S)
title ('column elimination tree')
```

Next, try this, which evaluates the quality of several
ordering methods with a sparse matrix from a chemical
process simulation problem:

```
load west0479 ;
A = west0479 ;
try_lu (A)
try_lu (A, @colperm)
try_lu (A, @symrcm, 1)
try_lu (A, @colmmd)
try_lu (A, @colamd)
```

See how much sparsity helped by trying this (the flop
count will be wrong, though):

```
try_lu (full (A))
```

13.5 Visualizing matrices

The previous section gave an example of how to use spy
to plot the nonzero pattern of a sparse matrix. spy can
also be used on full matrices. It is useful for matrix
expressions coming from relational operators. Try this,
for example (see Chapter 7 for the ddom function):

```
A = [
-1   2   3  -4
 0   2  -1   0
 1   2   9   1
-3   4   1   1]
C = ddom (A)
figure (2)
spy (A ~= C)
spy (A > 2)
```

What you see is a picture of where A and C differ, and another picture of which entries of A are greater than 2.

14. The Symbolic Math Toolbox

The Symbolic Math Toolbox, which utilizes the Maple V kernel as its computer algebra engine, lets you perform symbolic computation from within MATLAB. Under this configuration, MATLAB's numeric and graphic environment is merged with Maple's symbolic computation capabilities. The toolbox M-files that access these symbolic capabilities have names and syntax that will be natural for the MATLAB user. Key features of the Symbolic Math Toolbox are included in the Student Version of MATLAB. Since the Symbolic Math Toolbox is not part of the Professional Version of MATLAB, it may not be installed on your system, in which case this Chapter will not apply.

Many of the functions in the Symbolic Math Toolbox have the same names as their numeric counterparts. MATLAB selects the correct one depending on the type of inputs to the function. Typing `help eig` and `help sym/eig` displays the help for the numeric eigenvalue function and its symbolic counterpart, respectively.

14.1 Symbolic variables

You can declare a variable as symbolic with the `syms` statement. For example,

    syms x

creates a symbolic variable x. The statement:

    syms x real

declares to Maple that x is a symbolic variable with no imaginary part. Maple has its own workspace. The statements `clear` or `clear x` do **not** undo this declaration, because it clears MATLAB's variable x but not Maple's variable s. Use `syms x unreal`, which declares to Maple that x may now have a nonzero imaginary part. The `clear all` statement clears all variables in both MATLAB and Maple, and thus also resets the `real` or `unreal` status of x. You can also assert to Maple that x is always positive, with `syms x positive`.

Symbolic variables can be constructed from existing numeric variables using the `sym` function. Try:

```
z = 1/10
a = sym (z)
y = rand (1)
b = sym (y, 'd')
```

although a better way to create a is:

```
a = sym ('1/10')
```

The `syms` command and `sym` function have many more options. See `help syms` and `help sym`.

14.2 Calculus

The function `diff` computes the symbolic derivative of a function defined by a symbolic expression. First, to define a symbolic expression, you should create symbolic variables and then proceed to build an expression as you would mathematically. For example,

```
syms x
f = x^2 * exp (x)
diff (f)
```

creates a symbolic variable x, builds the symbolic
expression $f = x^2 e^x$, and returns the symbolic derivative of
f with respect to x: 2*x*exp(x)+x^2*exp(x) in
MATLAB notation. Try it.

Next,

```
syms t
diff (sin (pi*t))
```

returns the derivative of $\sin(\pi t)$, as a function of t.

Partial derivatives can also be computed. Try the
following:

```
syms x y
g = x*y + x^2
diff (g)          % computes ∂g/∂x
diff (g, x)       % also ∂g/∂x
diff (g, y)       % ∂g/∂y
```

To permit omission of the second argument for functions
such as the above, MATLAB chooses a default symbolic
variable for the symbolic expression. The findsym
function returns MATLAB's choice. Its rule is, roughly,
to choose that lower case letter, other than i and j, nearest
x in the alphabet.

You can, of course, override the default choice as shown
above. Try, for example,

```
syms x1 x2 theta
F = x * (x1*x2 + x1 - 2)
```

```
diff (F)                    % ∂F/∂x
diff (F, x1)                % ∂F/∂x1
diff (F, x2)                % ∂F/∂x2
G = cos (theta*x)
diff (G, theta)             % ∂G/∂theta
```

The second derivative, for example, can be obtained by the command:

```
diff (sin (2*x), x, 2)
```

With a numeric argument, diff is the difference operator of basic MATLAB, which can be used to numerically approximate the derivative of a function. See help diff for the numeric function, and help sym/diff for the symbolic derivative function.

The function int attempts to compute the indefinite integral (antiderivative) of a function defined by a symbolic expression. Try, for example,

```
syms a b t x y z theta
int (sin (a*t + b))
int (sin (a*theta + b), theta)
int (x*y^2 + y*z, y)
int (x^2 * sin (x))
```

Note that, as with diff, when the second argument of int is omitted, the default symbolic variable (as selected by findsym) is chosen as the variable of integration.

In some instances, int will be unable to give a result in terms of elementary functions. Consider, for example,

```
int (exp (-x^2))
int (sqrt (1 + x^3))
```

77

In the first case the result is given in terms of the error function `erf`, whereas in the second, the result is given in terms of `EllipticF`, a function defined by an integral.

The function `pretty` will display a symbolic expression in an easier-to-read form resembling typeset mathematics (see `latex`, `ccode`, and `fortran` for other formats). Try, for example,

```
syms x a b
f = x/(a*x+b)
pretty (f)
g = int (f)
pretty (g)
latex (g)
ccode (g)
fortran (g)
int (g)
pretty (ans)
```

Definite integrals can also be computed by using additional input arguments. Try, for example,

```
int (sin (x), 0, pi)
int (sin (theta), theta, 0, pi)
```

In the first case, the default symbolic variable x was used as the variable of integration to compute:

$$\int_0^\pi \sin x \, dx$$

whereas in the second `theta` was chosen. Other definite integrals you can try are:

```
int (x^5, 1, 2)
int (log (x), 1, 4)
```

```
int (x * exp (x), 0, 2)
int (exp (-x^2), 0, inf)
```

It is important to realize that the results returned are
symbolic expressions, not numeric ones. The function
double will convert these into MATLAB floating-point
numbers, if desired. For example, the result returned by
the first integral above is 21/2. Entering double(ans)
then returns the MATLAB numeric result 10.5000.

Alternatively, you can use the function vpa (variable
precision arithmetic; see Section 14.3) to convert the
expression into a symbolic number of arbitrary precision.
For example,

```
int (exp (-x^2), 0, inf)
```

gives the result:

```
1/2*pi^(1/2)
```

Then the statement:

```
vpa (ans, 25)
```

symbolically gives the result to 25 significant digits:

```
.8862269254527580136490835
```

You may wish to contrast these techniques with the
MATLAB numerical integration functions quad and
quad8.

The limit function is used to compute the symbolic
limits of various expressions. For example,

```
syms h n x
limit ((1 + x/n)^n, n, inf)
```

computes the limit of $(1 + x/n)^n$ as $n \rightarrow \infty$. You should also try:

```
limit (sin (x), x, 0)
limit ((sin(x+h)-sin(x))/h, h, 0)
```

The `taylor` function computes the Maclaurin and Taylor series of symbolic expressions. For example,

```
taylor (cos (x) + sin (x))
```

returns the 5th order Maclaurin polynomial approximating $\cos(x) + \sin(x)$. The command,

```
taylor (cos (x^2), 8, x, pi)
```

returns the 8th degree Taylor approximation to $\cos(x^2)$ centered at the point $x_0 = \pi$.

14.3 Variable precision arithmetic

Three kinds of arithmetic operations are available:

numeric	MATLAB's floating-point arithmetic
rational	Maple's exact symbolic arithmetic
VPA	Maple's variable precision arithmetic

One can obtain exact rational results with, for example,

```
s = simple (sym ('13/17 + 17/23'))
```

You are already familiar with numeric computations. For example, with `format long`,

```
pi*log(2)
```

gives the numeric result:

```
2.17758609030360
```

MATLAB's numeric computations are done in approximately 16 decimal digit floating-point arithmetic. With vpa, you can obtain results to arbitrary precision, within the limitations of time and memory. For example, try:

```
vpa ('pi * log (2)')
vpa ('pi * log (2)', 50)
```

The default precision for vpa is 32. Hence, the first result is accurate to 32 digits, whereas the second is accurate to the specified 50 digits.[5] The default precision can be changed with the function digits. While the rational and VPA computations can be more accurate, they are in general slower than numeric computations.

If you pass an expression to vpa, MATLAB will evaluate it numerically first, unless it is a symbolic expression or placed in quotes. Compare your results, above, with:

```
vpa (pi * log (2))
```

which is accurate to only about 16 digits (even though 32 digits are displayed). This is a common mistake with the use of vpa and the Symbolic Math Toolbox in general.

---

[5] Ludolf van Ceulen (1540-1610) calculated π to 36 digits. The Symbolic Math Toolbox can quite easily compute π to 10,000 digits or more. Try vpa('pi',10000).

14.4 Numeric evaluation

Once you have a symbolic expression, you can evaluate it
numerically with the `eval` function. Try:

```
syms x
F = x^2 * sin (x)
G = diff (F)
H = vectorize (G)
x = 0:.1:1
eval (H)
```

The `vectorize` function allows H to be evaluated with a
vector x. Also try:

```
syms x y
S = x^y
x = 3
eval (S)
y = 2
eval (S)
```

The `eval` function returns a symbolic expression unless
all of the variables are numeric.

14.5 Algebraic simplification

Convenient algebraic manipulations of symbolic
expressions are available.

The function `expand` distributes products over sums and
applies other identities, whereas `factor` attempts to do
the reverse. The function `collect` views a symbolic
expression as a polynomial in its symbolic variable
(which may be specified) and collects all terms with the
same power of the variable. To explore these capabilities,
try the following:

```
syms a b x y z
expand ((a + b)^5)
factor (ans)
expand (exp (x + y))
expand (sin (x + 2*y))
factor (x^6 - 1)
collect (x * (x * (x + 3) + 5) + 1)
horner (ans)
collect ((x + y + z)*(x - y - z))
collect ((x + y + z)*(x - y - z), y)
collect ((x + y + z)*(x - y - z), z)
diff (x^3 * exp (x))
factor (ans)
```

The powerful function simplify applies many identities in an attempt to reduce a symbolic expression to a simple form. Try, for example,

```
simplify (sin(x)^2 + cos(x)^2)
simplify (exp (5*log(x) + 1))
d = diff ((x^2 + 1)/(x^2 - 1))
simplify (d)
```

The alternate function simple computes several simplifications and chooses the shortest of them. It often gives better results on expressions involving trigonometric functions. Try the following commands:

```
simplify(cos(x) + (-sin(x)^2)^(1/2))
simple   (cos(x) + (-sin(x)^2)^(1/2))
simplify((1/x^3+6/x^2+12/x+8)^(1/3))
simple   ((1/x^3+6/x^2+12/x+8)^(1/3))
```

The function subs replaces all occurrences of the symbolic variable in an expression by a specified second expression. This corresponds to composition of two functions. Try, for example,

```
syms x s t
subs (sin(x), x, pi/3)
subs (sin(x), x, sym (pi)/3)
double (ans)
subs (g*t^2/2, t, sqrt(2*s))
subs (sqrt(1-x^2), x, cos(x))
subs (sqrt(1-x^2), 1-x^2, cos(x))
```

The general idea is that in the statement
subs(expr,old,new) the third argument (new)
replaces the second argument (old) in the first argument
(expr). Compare the first two examples above. The
result is numeric if all variables in the expression are
substituted with numeric values.

The function factor can also be applied to an integer
argument to compute the prime factorization of the
integer. Try, for example,

```
factor (sym ('4248'))
factor (sym ('4549319348693'))
factor (sym ('4549319348597'))
```

14.6 Graphs of functions

The MATLAB function fplot (see Section 10.3)
provides a tool to conveniently plot the graph of a
function. Since it is, however, the name or handle of the
function to be plotted that is passed to fplot, the
function must first be defined in an M-file (or else be a
built-in function or inline function).

In the Symbolic Math Toolbox, ezplot lets you plot the
graph of a function directly from its defining symbolic
expression. For example, try:

```
syms t x
ezplot (sin (2*x))
```

```
ezplot (t + 3*sin(t))
ezplot (2*x/(x^2 - 1))
ezplot (1/(1 + 30*exp(-x)))
```

By default, the x-domain is [-2*pi, 2*pi]. This can
be overridden by a second input variable, as with:

```
ezplot(x*sin(1/x), [-.2 .2])
```

You will often need to specify the x-domain and y-
domain to zoom in on the relevant portion of the graph.
Compare, for example,

```
ezplot (x*exp(-x))
ezplot (x*exp(-x), [-1 4])
```

ezplot attempts to make a reasonable choice for the y-
axis. With the last figure, select Edit ▶ Axes
Properties in the Figure window and modify the y-axis
to start at -3, and click OK. Changing the x-axis in the
Property Editor does not cause the function to be
reevaluated, however.

Entering the command funtool (no input arguments)
brings up three graphic figures, two of which will display
graphs of functions and one containing a control panel.
This function calculator lets you manipulate functions and
their graphs for pedagogical demonstrations. Type help
funtool for details.

14.7 Symbolic matrix operations

This toolbox lets you represent matrices in symbolic form
as well as MATLAB's numeric form. Given the numeric
matrix:

```
a = magic (3)
```

the function `sym(a)` converts `a` to the symbolic matrix.
Try:

```
A = sym (a)
```

The result is:

```
[8, 1, 6]
[3, 5, 7]
[4, 9, 2]
```

The function `numeric(A)` converts the symbolic matrix
back to a numeric one.

Symbolic matrices can also be generated by `sym`. Try, for
example,

```
syms a b s
K = [a + b, a - b ; b - a, a + b]
G = [cos(s), sin(s); -sin(s), cos(s)]
```

Here `G` is a symbolic Givens rotation matrix.

Algebraic matrix operations with symbolic matrices are
computed as you would in MATLAB.

K+G	matrix addition
K-G	matrix subtraction
K*G	matrix multiplication
inv(G)	matrix inversion
K/G	right matrix division
K\G	left matrix division
G^2	power
G.'	transpose
G'	conjugate transpose (Hermitian)

These operations are illustrated by the following, which use the matrices K and G generated above:

```
L = K^2
collect (L)
factor (L)
diff (L, a)
int (K, a)
J = K/G
simplify (J*G)
simplify (G*(G.'))
```

Note that the initial result of the basic operations may not be in the form desired for your application; so it may require further processing with simplify, collect, factor, or expand. These functions, as well as diff and int, act entry-wise on a symbolic matrix.

14.8 Symbolic linear algebraic functions

The primary symbolic matrix functions are:

det	determinant
.'	transpose
'	Hermitian (conjugate transpose)
inv	inverse
null	basis for nullspace
colspace	basis for column space
eig	eigenvalues and eigenvectors
poly	characteristic polynomial
svd	singular value decomposition
jordan	Jordan canonical form

These functions will take either symbolic or numeric arguments.

Computations with symbolic rational matrices can be carried out exactly. Try, for example,

```
c = floor (10*rand(4))
D = sym (c)
A = inv (D)
inv (A)
det (A)
b = ones (1,4)
x = b/A
x*A
A^3
```

These functions can, of course, be applied to general symbolic matrices. For the matrices K and G defined in the previous section, try:

```
inv (K)
simplify (inv (G))
p = poly (G)
simplify (p)
factor (p)
X = solve (p)
for j = 1:4
    X = simple (X)
end
pretty (X)
e = eig (G)
for j = 1:4
    e = simple (e)
end
pretty (e)
y = svd (G)
for j = 1:4
    y = simple (y)
end
pretty (y)
syms s real
r = svd (G)
r = simple (r)
```

88

```
pretty (r)
syms s unreal
```

See Section 14.9 on the solve function.

A typical exercise in a linear algebra course is to determine those values of t so that, say,

```
A = [t 1 0 ; 1 t 1 ; 0 1 t]
```

is singular. The following simple computation:

```
syms t
A = [t 1 0 ; 1 t 1 ; 0 1 t]
p = det (A)
solve (p)
```

shows that this occurs for t = 0, $\sqrt{2}$, and $\sqrt{-2}$.

The function eig attempts to compute the eigenvalues and eigenvectors in an exact closed form. Try, for example,

```
for n = 4:6
    A = sym (magic (n))
    [V, D] = eig (A)
end
```

Except in special cases, however, the result is usually too complicated to be useful. Try, for example, executing:

```
A = sym (floor (10 * rand (3)))
[V, D] = eig (A)
```

a few times. For this reason, it is usually more efficient to do the computation in variable-precision arithmetic, as is illustrated by:

```
A = vpa (floor (10 * rand(3)))
[V, D] = eig (A)
```

The comments above regarding `eig` apply as well to the computation of the singular values of a matrix by `svd`, as can be observed by repeating some of the computations above using `svd` instead of `eig`.

14.9 Solving algebraic equations

For a symbolic expression S, the statement `solve(s)` will attempt to find the values of the symbolic variable for which the symbolic expression is zero. If an exact symbolic solution is indeed found, you can convert it to a floating-point solution, if desired. If an exact symbolic solution cannot be found, then a variable precision one is computed. Moreover, if you have an expression that contains several symbolic variables, you can solve for a particular variable by including it as an input argument in `solve`. The inputs to `solve` can be quoted strings or symbolic expressions.

Try these symbolic expressions, for example:

```
syms x y z
X = solve (cos(x) + tan(x))
pretty (X)
double (X)
vpa (X)
Y = solve (cos(x) - x)
Z = solve (x^2 + 2*x - 1)
pretty (Z)
a = solve (x^2 + y^2 + z^2 + x*y*z)
pretty (a)
b = solve (x^2 + y^2 + z^2 + x*y*z, y)
pretty (b)
```

The result a is a solution in the variable x, and b is a solution in y. To solve an equation whose right-hand side is not 0, use a quoted string. Some examples are:

```
X = solve ('log(x) = x - 2')
vpa (X)
X = solve ('2^x = x + 2')
vpa (X)
```

This solves for the variable a:

```
A = solve ('1 + (a+b)/(a-b) = b', 'a')
```

and this solves the same equation for b:

```
f = solve ('1 + (a+b)/(a-b) = b', 'b')
```

The function solve can also compute the solutions of systems of general algebraic equations. To solve, for example, the nonlinear system below, it is convenient to first express the equations as strings.

```
S1 = 'x^2 + y^2 + z^2 = 2'
S2 = 'x + y = 1'
S3 = 'y + z = 1'
```

The solutions are then computed by:

```
[X, Y, Z] = solve (S1, S2, S3)
```

If you alter S2 to:

```
S2 = 'x + y + z = 1'
```

then the solution computed by:

```
[X, Y, Z] = solve (S1, S2, S3)
```

will be given in terms of square roots.

The `solve` function can take quoted strings or symbolic expressions as input arguments, but you cannot mix the two types of inputs.

14.10 Solving differential equations

The function `dsolve` attempts to solve ordinary differential equations. The symbolic differential operator is D, so that:

```
Y = dsolve ('Dy = x^2*y', 'x')
```

produces the solution `C1*exp(1/3*x^3)` to the differential equation $y' = x^2 y$. The solution to an initial value problem can be computed by adding a second symbolic expression giving the initial condition.

```
Y = dsolve ('Dy = x^2*y', 'y(0)=4', 'x')
```

Notice that in both examples above, the final input argument, `'x'`, is the independent variable of the differential equation. If no independent variable is supplied to `dsolve`, then it is assumed to be t. The higher order symbolic differential operators D2, D3, ... can be used to solve higher order equations. Explore the following:

```
dsolve ('D2y + y = 0')
dsolve ('D2y + y = x^2', 'x')
dsolve ('D2y + y = x^2', ...
    'y(0) = 4', 'Dy(0) = 1', 'x')
dsolve ('D2y - Dy = 2*y')
dsolve ('D2y + 6*Dy = 13*y')
Y = dsolve ('D2y + 6*Dy + 13*y =
cos(t)')
Y = simple (Y)
```

```
dsolve ('D3y - 3*Dy = 2*y')
pretty (ans)
```

Systems of differential equations can also be solved. For example,

```
E1 = 'Dx = -2*x + y'
E2 = 'Dy = x - 2*y + z'
E3 = 'Dz = y - 2*z'
```

The solutions are then computed with:

```
[x, y, z] = dsolve (E1, E2, E3)
pretty (x)
pretty (y)
pretty (z)
```

You can explore further details with help dsolve.

14.11 Further Maple access

The following features are not available in the Student Version of MATLAB.

Over 50 special functions of classical applied mathematics are available in the Symbolic Math Toolbox. Enter help mfunlist to see a list of them. These functions can be accessed with the function mfun, for which you are referred to help mfun for further details. The maple function allows you to use expressions and programming constructs in Maple's native language, which gives you full access to Maple's functionality. See help maple, or mhelp *topic*, which displays Maple's help text for the specified topic. The Extended Symbolic Math Toolbox provides access to a number of Maple's specialized libraries of procedures. It also provides for use of Maple programming features.

15. Help topics

There are many MATLAB functions and features that cannot be included in this Primer. Listed in the following tables are some of the MATLAB functions and operators, grouped by subject area.[6] You can browse through these lists and use the online help facility, or consult the online documents *MATLAB Functions: Volumes 1 through 3* for more detailed information on the functions, operators, and special characters.

Typing help at the MATLAB command prompt will provide a listing of the major MATLAB directories, similar to the following table. Typing help *topic*, where *topic* is an entry in the left column of the table, will display a description of the topic. For example, help general will display on your Command window a plain text version of Section 15.1. Typing help ops will display Section 15.2, starting on page 99, and so on.

Each topic is discussed in a single subsection. The page number for each subsection is also listed in the following table.

---

[6] Source: MATLAB 6.1 help command, Release R12.1.

Help topics		page
general	General purpose commands	96
ops	Operators and special characters	99
lang	Programming language constructs	101
elmat	Elementary matrices and matrix manipulation	104
elfun	Elementary math functions	106
specfun	Specialized math functions	108
matfun	Matrix functions–numerical linear algebra	110
datafun	Data analysis and Fourier transforms	112
audio	Audio support	113
polyfun	Interpolation and polynomials	115
funfun	Function functions and ODE solvers	116
sparfun	Sparse matrices	119
graph2d	Two-dimensional graphs	121
graph3d	Three-dimensional graphs	122
specgraph	Specialized graphs	125
graphics	Handle Graphics	129
uitools	Graphical user interface tools	131
strfun	Character strings	134
iofun	File input/output	136
timefun	Time and dates	139
datatypes	Data types and structures	140
verctrl	Version control	143
winfun	Microsoft Windows Interface Files	144
demos	Examples and demonstrations	144
local	Preferences	144
symbolic	Symbolic Math Toolbox	145

15.1 General

`help general`

General information	
`helpbrowser`	Bring up the help browser
`doc`	Complete online help, displayed in the help browser (`helpdesk` in Version 6.0)
`help`	M-file help, displayed in the Command window
`helpwin`	M-file help, displayed in the help browser
`lookfor`	Search all M-files for keyword
`syntax`	Help on MATLAB command syntax
`support`	Open MathWorks technical support web page
`demo`	Run demonstrations
`ver`	MATLAB, Simulink, and toolbox version information
`version`	MATLAB version information
`whatsnew`	Access release notes

Managing the workspace	
`who`	List current variables
`whos`	List current variables, long form
`workspace`	Display Workspace window
`clear`	Clear variables and functions from memory
`pack`	Consolidate workspace memory
`load`	Load workspace variables from disk
`save`	Save workspace variables to disk
`quit`	Quit MATLAB session

Managing commands and functions

what	List MATLAB-specific files in directory
type	List M-file
edit	Edit M-file
open	Open files by extension
which	Locate functions and files
pcode	Create pre-parsed pseudo-code file (P-file)
inmem	List functions in memory
mex	Compile MEX-function

Managing the search path

path	Get/set search path
addpath	Add directory to search path
rmpath	Remove directory from search path
pathtool	Modify search path
rehash	Refresh function and file system caches
import	Import Java packages into the current scope

Controlling the Command window

echo	Echo commands in M-files
more	Control paged output in Command window
diary	Save text of MATLAB session
format	Set output format
beep	Produce beep sound

Operating system commands

cd	Change current working directory
copyfile	Copy a file
pwd	Show (print) current working directory
dir	List directory
delete	Delete file

(continued on next page)

Operating system commands (continued)

`getenv`	Get environment variable
`mkdir`	Make directory
`!`	Execute operating system command
`dos`	Execute DOS command and return result
`unix`	Execute Unix command and return result
`system`	Execute system command and return result
`web`	Open web browser on site or files
`computer`	Computer type
`isunix`	True for the Unix version of MATLAB
`ispc`	True for the Windows version of MATLAB

Debugging M-files

`debug`	List debugging commands
`dbstop`	Set breakpoint
`dbclear`	Remove breakpoint
`dbcont`	Continue execution
`dbdown`	Change local workspace context
`dbstack`	Display function call stack
`dbstatus`	List all breakpoints
`dbstep`	Execute one or more lines
`dbtype`	List M-file with line numbers
`dbup`	Change local workspace context
`dbquit`	Quit debug mode
`dbmex`	Debug MEX-files (Unix only)

Profiling M-files

`profile`	Profile function execution time
`profreport`	Generate profile report

Locate dependent functions of an M-file	
depfun	Locate dependent functions of an M-file
depdir	Locate dependent directories of an M-file
inmem	List functions in memory

15.2 Operators and special characters

help ops

Arithmetic operators (help arith, help slash)		
plus	Plus	+
uplus	Unary plus	+
minus	Minus	−
uminus	Unary minus	−
mtimes	Matrix multiply	*
times	Array multiply	.*
mpower	Matrix power	^
power	Array power	.^
mldivide	left matrix divide	\
mrdivide	right matrix divide	/
ldivide	Left array divide	.\
rdivide	Right array divide	./
kron	Kronecker tensor product	kron

Relational operators (help relop)		
eq	Equal	==
ne	Not equal	~=
lt	Less than	<
gt	Greater than	>
le	Less than or equal	<=
ge	Greater than or equal	>=

Logical operators		
and	Logical AND	&
or	Logical OR	|
not	Logical NOT	~
xor	Logical EXCLUSIVE OR	
any	True if any element of vector is nonzero	
all	True if all elements of vector are nonzero	

Special characters		
colon	Colon	:
paren	Parentheses and subscripting	( )
paren	Brackets	[ ]
paren	Braces and subscripting	{ }
punct	Function handle creation	@
punct	Decimal point	.
punct	Structure field access	.
punct	Parent directory	..
punct	Continuation	...
punct	Separator	,
punct	Semicolon	;
punct	Comment	%
punct	Invoke operating system command	!
punct	Assignment	=
punct	Quote	'
transpose	Transpose	.'
ctranspose	Complex conjugate transpose	'
horzcat	Horizontal concatenation	[,]
vertcat	Vertical concatenation	[;]
subsasgn	Subscripted assignment	( ) { }
subsref	Subscripted reference	( ) { }
subsindex	Subscript index	

Bitwise operators	
bitand	Bit-wise AND
bitcmp	Complement bits
bitor	Bit-wise OR
bitmax	Maximum floating-point integer
bitxor	Bit-wise EXCLUSIVE OR
bitset	Set bit
bitget	Get bit
bitshift	Bit-wise shift

Set operators	
union	Set union
unique	Set unique
intersect	Set intersection
setdiff	Set difference
setxor	Set exclusive-or
ismember	True for set member

15.3 Programming language constructs

help lang

Control flow	
if	Conditionally execute statements
else	if statement condition
elseif	if statement condition
end	Terminate scope of for, while, switch, try and if statements
for	Repeat statements a specific number of times
while	Repeat statements an indefinite number of times
break	Terminate execution of while or for loop

(continued on next page)

Control flow (continued)

continue	Pass control to the next iteration of for or while loop
switch	Switch among several cases based on expression
case	switch statement case
otherwise	Default switch statement case
try	Begin try block
catch	Begin catch block
return	Return to invoking function

Evaluation and execution

eval	Execute string with MATLAB expression
evalc	Evaluate MATLAB expression with capture
feval	Execute function specified by string
evalin	Evaluate expression in workspace
builtin	Execute built-in function from overloaded method
assignin	Assign variable in workspace
run	Run script

Scripts, functions, and variables

script	About MATLAB scripts and M-files
function	Add new function
global	Define global variable
persistent	Define persistent variable
mfilename	Name of currently executing M-file
lists	Comma separated lists
exist	Check if variables or functions are defined
isglobal	True for global variables
mlock	Prevent M-file from being cleared

(continued on next page)

Scripts, functions, and variables (cont.)

munlock	Allow M-file to be cleared
mislocked	True if M-file cannot be cleared
precedence	Operator precedence in MATLAB
isvarname	Check for a valid variable name
iskeyword	Check if input is a keyword

Argument handling

nargchk	Validate number of input arguments
nargoutchk	Validate number of output arguments
nargin	Number of function input arguments
nargout	Number of function output arguments
varargin	Variable length input argument list
varargout	Variable length output argument list
inputname	Input argument name

Message display

error	Display error message and abort function
warning	Display warning message
lasterr	Last error message
lastwarn	Last warning message
disp	Display an array
display	Overloaded function to display an array
fprintf	Display formatted message
sprintf	Write formatted data to a string

Interactive input

input	Prompt for user input
keyboard	Invoke keyboard from M-file
pause	Wait for user response
uimenu	Create user interface menu
uicontrol	Create user interface control

15.4 Elementary matrices and matrix manipulation

`help elmat`

Elementary matrices	
`zeros`	Zeros array
`ones`	Ones array
`eye`	Identity matrix
`repmat`	Replicate and tile array
`rand`	Uniformly distributed random numbers
`randn`	Normally distributed random numbers
`linspace`	Linearly spaced vector
`logspace`	Logarithmically spaced vector
`freqspace`	Frequency spacing for frequency response
`meshgrid`	x and y arrays for 3-D plots
`:`	Regularly spaced vector and index into matrix

Basic array information	
`size`	Size of matrix
`length`	Length of vector
`ndims`	Number of dimensions
`numel`	Number of elements
`disp`	Display matrix or text
`isempty`	True for empty matrix
`isequal`	True if arrays are identical
`isnumeric`	True for numeric arrays
`islogical`	True for logical array
`logical`	Convert numeric values to logical

Matrix manipulation

reshape	Change size
diag	Diagonal matrices; diagonals of matrix
blkdiag	Block diagonal concatenation
tril	Extract lower triangular part
triu	Extract upper triangular part
fliplr	Flip matrix in left/right direction
flipud	Flip matrix in up/down direction
flipdim	Flip matrix along specified dimension
rot90	Rotate matrix 90 degrees
:	Regularly spaced vector and index into matrix
find	Find indices of nonzero elements
end	Last index
sub2ind	Linear index from multiple subscripts
ind2sub	Multiple subscripts from linear index

Special variables and constants

ans	Most recent answer
eps	Floating-point relative accuracy
realmax	Largest positive floating-point number
realmin	Smallest positive floating-point number
pi	3.1415926535897...
i, j	Imaginary unit
inf	Infinity
NaN	Not-a-Number
isnan	True for Not-a-Number
isinf	True for infinite elements
isfinite	True for finite elements
why	Succinct answer

Specialized matrices	
compan	Companion matrix
gallery	Higham test matrices
hadamard	Hadamard matrix
hankel	Hankel matrix
hilb	Hilbert matrix
invhilb	Inverse Hilbert matrix
magic	Magic square
pascal	Pascal matrix
rosser	Classic symmetric eigenvalue test problem
toeplitz	Toeplitz matrix
vander	Vandermonde matrix
wilkinson	Wilkinson's eigenvalue test matrix

15.5 Elementary math functions

help elfun

Trigonometric	
sin	Sine
sinh	Hyperbolic sine
asin	Inverse sine
asinh	Inverse hyperbolic sine
cos	Cosine
cosh	Hyperbolic cosine
acos	Inverse cosine
acosh	Inverse hyperbolic cosine
tan	Tangent
tanh	Hyperbolic tangent
atan	Inverse tangent
atan2	Four quadrant inverse tangent
atanh	Inverse hyperbolic tangent
sec	Secant
sech	Hyperbolic secant

(continued on next page)

106

Trigonometric (continued)

`asec`	Inverse secant
`asech`	Inverse hyperbolic secant
`csc`	Cosecant
`csch`	Hyperbolic cosecant
`acsc`	Inverse cosecant
`acsch`	Inverse hyperbolic cosecant
`cot`	Cotangent
`coth`	Hyperbolic cotangent
`acot`	Inverse cotangent
`acoth`	Inverse hyperbolic cotangent

Exponential

`exp`	Exponential
`log`	Natural logarithm
`log10`	Common (base 10) logarithm
`log2`	Base 2 logarithm and dissect floating-point number
`pow2`	Base 2 power and scale floating-point number
`sqrt`	Square root
`nextpow2`	Next higher power of 2

Complex

`abs`	Absolute value
`angle`	Phase angle
`complex`	Construct complex data from real and imaginary parts
`conj`	Complex conjugate
`imag`	Complex imaginary part
`real`	Complex real part
`unwrap`	Unwrap phase angle
`isreal`	True for real array
`cplxpair`	Sort numbers into complex conjugate pairs

Rounding and remainder	
fix	Round towards zero
floor	Round towards minus infinity
ceil	Round towards plus infinity
round	Round towards nearest integer
mod	Modulus (signed remainder after division)
rem	Remainder after division
sign	Signum

15.6 Specialized math functions

help specfun

Specialized math functions	
airy	Airy functions
besselj	Bessel function of the first kind
bessely	Bessel function of the second kind
besselh	Bessel function of the third kind (Hankel function)
besseli	Modified Bessel function of the first kind
besselk	Modified Bessel function of the second kind
beta	Beta function
betainc	Incomplete beta function
betaln	Logarithm of beta function
ellipj	Jacobi elliptic functions
ellipke	Complete elliptic integral
erf	Error function
erfc	Complementary error function
erfcx	Scaled complementary error function
erfinv	Inverse error function
expint	Exponential integral function
gamma	Gamma function

(continued on next page)

Specialized math functions (continued)

gammainc	Incomplete gamma function
gammaln	Logarithm of gamma function
legendre	Associated Legendre function
cross	Vector cross product
dot	Vector dot product

Number theoretic functions

factor	Prime factors
isprime	True for prime numbers
primes	Generate list of prime numbers
gcd	Greatest common divisor
lcm	Least common multiple
rat	Rational approximation
rats	Rational output
perms	All possible permutations
nchoosek	All combinations of N elements taken K at a time
factorial	Factorial function

Coordinate transforms

cart2sph	Transform Cartesian to spherical coordinates
cart2pol	Transform Cartesian to polar coordinates
pol2cart	Transform polar to Cartesian coordinates
sph2cart	Transform spherical to Cartesian coordinates
hsv2rgb	Convert hue-saturation-value colors to red-green-blue
rgb2hsv	Convert red-green-blue colors to hue-saturation-value

15.7 Matrix functions — numerical linear algebra

`help matfun`

Matrix analysis	
`norm`	Matrix or vector norm
`normest`	Estimate the matrix 2-norm
`rank`	Matrix rank
`det`	Determinant
`trace`	Sum of diagonal elements
`null`	Null space
`orth`	Orthogonalization
`rref`	Reduced row echelon form
`subspace`	Angle between two subspaces

Linear equations	
`\ and /`	Linear equation solution; use `help slash`
`inv`	Matrix inverse
`rcond`	LAPACK reciprocal condition estimator
`cond`	Condition number with respect to inversion
`condest`	1-norm condition number estimate
`normest1`	1-norm estimate
`chol`	Cholesky factorization
`cholinc`	Incomplete Cholesky factorization
`lu`	LU factorization
`luinc`	Incomplete LU factorization
`qr`	Orthogonal-triangular decomposition
`lsqnonneg`	Linear least squares with nonnegativity constraints
`pinv`	Pseudoinverse
`lscov`	Least squares with known covariance

Eigenvalues and singular values

eig	Eigenvalues and eigenvectors
svd	Singular value decomposition
gsvd	Generalized singular value decomposition
eigs	A few eigenvalues
svds	A few singular values
poly	Characteristic polynomial
polyeig	Polynomial eigenvalue problem
condeig	Condition number with respect to eigenvalues
hess	Hessenberg form
qz	QZ factorization for generalized eigenvalues
schur	Schur decomposition

Matrix functions

expm	Matrix exponential
logm	Matrix logarithm
sqrtm	Matrix square root
funm	Evaluate general matrix function

Factorization utilities

qrdelete	Delete column from QR factorization
qrinsert	Insert column in QR factorization
rsf2csf	Real block diagonal form to complex diagonal form
cdf2rdf	Complex diagonal form to real block diagonal form
balance	Diagonal scaling to improve eigenvalue accuracy
planerot	Givens plane rotation
cholupdate	rank 1 update to Cholesky factorization
qrupdate	rank 1 update to QR factorization

15.8 Data analysis and Fourier transforms

`help datafun`

Basic operations	
max	Largest component
min	Smallest component
mean	Average or mean value
median	Median value
std	Standard deviation
var	Variance
sort	Sort in ascending order
sortrows	Sort rows in ascending order
sum	Sum of elements
prod	Product of elements
hist	Histogram
histc	Histogram count
trapz	Trapezoidal numerical integration
cumsum	Cumulative sum of elements
cumprod	Cumulative product of elements
cumtrapz	Cumulative trapezoidal numerical integration

Finite differences	
diff	Difference and approximate derivative
gradient	Approximate gradient
del2	Discrete Laplacian

Correlation	
corrcoef	Correlation coefficients
cov	Covariance matrix
subspace	Angle between subspaces

Filtering and convolution	
filter	One-dimensional digital filter
filter2	Two-dimensional digital filter
conv	Convolution and polynomial multiplication
conv2	Two-dimensional convolution
convn	N-dimensional convolution
deconv	Deconvolution and polynomial division
detrend	Linear trend removal

Fourier transforms	
fft	Discrete Fourier transform
fft2	2-D discrete Fourier transform
fftn	N-dimensional discrete Fourier transform
ifft	Inverse discrete Fourier transform
ifft2	2-D inverse discrete Fourier transform
ifftn	N-dimensional inverse discrete Fourier transform
fftshift	Shift zero-frequency component to center of spectrum
ifftshift	Inverse FFTSHIFT

15.9 Audio support

help audio

Audio input/output objects	
audioplayer	Windows audio player object
audiorecorder	Windows audio recorder object

Audio hardware drivers

sound	Play vector as sound
soundsc	Autoscale and play vector as sound
wavplay	Play sound using Windows audio output device
wavrecord	Record sound using Windows audio input device

Audio file import and export

auread	Read NeXT/SUN (.au) sound file
auwrite	Write NeXT/SUN (.au) sound file
wavread	Read Microsoft WAVE (.wav) sound file
wavwrite	Write Microsoft WAVE (.wav) sound file

Utilities

lin2mu	Convert linear signal to mu-law encoding
mu2lin	Convert mu-law encoding to linear signal

Example audio data (MAT files)

chirp	Frequency sweeps
gong	Gong
handel	Hallelujah chorus
laughter	Laughter from a crowd
splat	Chirp followed by a splat
train	Train whistle

15.10 Interpolation and polynomials

`help polyfun`

Data interpolation	
pchip	Piecewise cubic Hermite interpolating polynomial
interp1	1-D interpolation (table lookup)
interp1q	Quick 1-D linear interpolation
interpft	1-D interpolation using FFT method
interp2	2-D interpolation (table lookup)
interp3	3-D interpolation (table lookup)
interpn	N-D interpolation (table lookup)
griddata	Data gridding and surface fitting
griddata3	Data gridding and hyper-surface fitting for three-dimensional data
griddatan	Data gridding and hyper-surface fitting (dimension ≥ 2)

Spline interpolation	
spline	Cubic spline interpolation
ppval	Evaluate piecewise polynomial

Geometric analysis	
delaunay	Delaunay triangulation
delaunay3	3-D Delaunay tessellation
delaunayn	N-D Delaunay tessellation
dsearch	Search Delaunay triangulation for nearest point
dsearchn	Search N-D Delaunay tessellation for nearest point
tsearch	Closest triangle search
tsearchn	N-D closest triangle search
convhull	Convex hull
convhulln	N-D convex hull
voronoi	Voronoi diagram

(continued on next page)

Geometric analysis (continued)	
voronoin	N-D Voronoi diagram
inpolygon	True for points inside polygonal region
rectint	Rectangle intersection area
polyarea	Area of polygon

Polynomials	
roots	Find polynomial roots
poly	Convert roots to polynomial
polyval	Evaluate polynomial
polyvalm	Evaluate polynomial with matrix argument
residue	Partial-fraction expansion (residues)
polyfit	Fit polynomial to data
polyder	Differentiate polynomial
polyint	Integrate polynomial analytically
conv	Multiply polynomials
deconv	Divide polynomials

15.11 Function functions and ODE solvers

help funfun

Optimization and root finding	
fminbnd	Scalar bounded nonlinear function minimization
fminsearch	Multidimensional unconstrained nonlinear minimization
fzero	Scalar nonlinear zero finding

Optimization option handling	
optimset	Create or alter optimization options structure
optimget	Get optimization parameters from options structure

Numerical integration (quadrature)

quad	Numerically evaluate integral, low order method
quadl	Numerically evaluate integral, higher order method
dblquad	Numerically evaluate double integral

Plotting

ezplot	Easy-to-use function plotter
ezplot3	Easy-to-use 3-D parametric curve plotter
ezpolar	Easy-to-use polar coordinate plotter
ezcontour	Easy-to-use contour plotter
ezcontourf	Easy-to-use filled contour plotter
ezmesh	Easy-to-use 3-D mesh plotter
ezmeshc	Easy-to-use mesh/contour plotter
ezsurf	Easy-to-use 3-D colored surface plotter
ezsurfc	Easy-to-use surf/contour plotter
fplot	Plot function

Inline function object

inline	Construct inline function object
argnames	Argument names
formula	Function formula
char	Convert inline object to char. array

Differential equation solvers

ode45	Solve non-stiff differential equations, medium order method
ode23	Solve non-stiff differential equations, low order method
ode113	Solve non-stiff differential equations, variable order method
ode23t	Solve moderately stiff ODEs and DAEs Index 1, trapezoidal rule

(continued on next page)

Differential equation solvers (continued)

ode15s	Solve stiff ODEs and DAEs Index 1, variable order method
ode23s	Solve stiff differential equations, low order method
ode23tb	Solve stiff differential equations, low order method

Boundary value problem solver for ODEs

bvp4c	Solve two-point boundary value problems for ODEs by collocation

1-D Partial differential equation solver

pdepe	Solve initial-boundary value problems for parabolic-elliptic PDEs

Option handling

odeset	Create/alter ODE options structure
odeget	Get ODE options parameters
bvpset	Create/alter BVP options structure
bvpget	Get BVP options parameters

Input and output functions

deval	Evaluates the solution of a differential equation problem (replaces bvpval)
odeplot	Time series ODE output function
odephas2	2-D phase plane ODE output function
odephas3	3-D phase plane ODE output function
odeprint	Command window printing ODE output function
bvpinit	Forms the initial guess for BVP4C
pdeval	Evaluates by interpolation the solution computed by PDEPE
odefile	MATLAB v5 ODE file syntax (obsolete)
bvpval	Evaluate solution (obsolete; use deval)

15.12 Sparse matrices

`help sparfun`

Elementary sparse matrices	
speye	Sparse identity matrix
sprand	Sparse uniformly distributed random matrix
sprandn	Sparse normally distributed random matrix
sprandsym	Sparse random symmetric matrix
spdiags	Sparse matrix formed from diagonals

Full to sparse conversion	
sparse	Create sparse matrix
full	Convert sparse matrix to full matrix
find	Find indices of nonzero elements
spconvert	Import from sparse matrix external format

Working with sparse matrices	
nnz	Number of nonzero matrix elements
nonzeros	Nonzero matrix elements
nzmax	Amount of storage allocated for nonzero matrix elements
spones	Replace nonzero sparse matrix elements with ones
spalloc	Allocate space for sparse matrix
issparse	True for sparse matrix
spfun	Apply function to nonzero matrix elements
spy	Visualize sparsity pattern

Reordering algorithms

`colamd`	Column approximate minimum degree permutation
`symamd`	Symmetric approximate minimum degree permutation
`colmmd`	Column minimum degree permutation
`symmmd`	Symmetric minimum degree permutation
`symrcm`	Symmetric reverse Cuthill-McKee permutation
`colperm`	Column permutation
`randperm`	Random permutation
`dmperm`	Dulmage-Mendelsohn permutation

Linear algebra

`eigs`	A few eigenvalues, using ARPACK
`svds`	A few singular values, using `eigs`
`luinc`	Incomplete LU factorization
`cholinc`	Incomplete Cholesky factorization
`normest`	Estimate the matrix 2-norm
`condest`	1-norm condition number estimate
`sprank`	Structural rank

Linear equations (iterative methods)

`pcg`	Preconditioned conjugate gradients method
`bicg`	Biconjugate gradients method
`bicgstab`	Biconjugate gradients stabilized method
`cgs`	Conjugate gradients squared method
`gmres`	Generalized minimum residual method
`minres`	Minimum residual method
`qmr`	Quasi-minimal residual method
`symmlq`	Symmetric LQ method

Operations on graphs (trees)	
treelayout	Lay out tree or forest
treeplot	Plot picture of tree
etree	Elimination tree
etreeplot	Plot elimination tree
gplot	Plot graph, as in "graph theory"

Miscellaneous	
symbfact	Symbolic factorization analysis
spparms	Set parameters for sparse matrix routines
spaugment	Form least squares augmented system

15.13 Two-dimensional graphs

help graph2d

Elementary x-y graphs	
plot	Linear plot
loglog	Log-log scale plot
semilogx	Semi-log scale plot
semilogy	Semi-log scale plot
polar	Polar coordinate plot
plotyy	Graphs with y tick labels on left & right

Axis control	
axis	Control axis scaling and appearance
zoom	Zoom in and out on a 2-D plot
grid	Grid lines
box	Axis box
hold	Hold current graph
axes	Create axes in arbitrary positions
subplot	Create axes in tiled positions

Graph annotation	
plotedit	Tools for editing and annotating plots
legend	Graph legend
title	Graph title
xlabel	x-axis label
ylabel	y-axis label
texlabel	Produces TeX format from a character string
text	Text annotation
gtext	Place text with mouse

Hard copy and printing	
print	Print graph or Simulink system; or save graph to M-file
printopt	Printer defaults
orient	Set paper orientation

15.14 Three-dimensional graphs

help graph3d

Elementary 3-D plots	
plot3	Plot lines and points in 3-D space
mesh	3-D mesh surface
surf	3-D colored surface
fill3	Filled 3-D polygons

Color control	
colormap	Color look-up table
caxis	Pseudocolor axis scaling
shading	Color shading mode
hidden	Mesh hidden line removal mode
brighten	Brighten or darken color map
colordef	Set color defaults
graymon	Set graphics defaults for grayscale monitors

Lighting

surfl	3-D shaded surface with lighting
lighting	Lighting mode
material	Material reflectance mode
specular	Specular reflectance
diffuse	Diffuse reflectance
surfnorm	Surface normals

Color maps

hsv	Hue-saturation-value color map
hot	Black-red-yellow-white color map
gray	Linear grayscale color map
bone	Grayscale with tinge of blue color map
copper	Linear copper-tone color map
pink	Pastel shades of pink color map
white	All-white color map
flag	Alternating red, white, blue, and black color map
lines	Color map with the line colors
colorcube	Enhanced color-cube color map
vga	Windows colormap for 16 colors
jet	Variant of HSV
prism	Prism color map
cool	Shades of cyan and magenta color map
autumn	Shades of red and yellow color map
spring	Shades of magenta and yellow color map
winter	Shades of blue and green color map
summer	Shades of green and yellow color map

Transparency

alpha	Transparency (alpha) mode
alphamap	Transparency (alpha) look-up table
alim	Transparency (alpha) scaling

Axis control

axis	Control axis scaling and appearance
zoom	Zoom in and out on a 2-D plot
grid	Grid lines
box	Axis box
hold	Hold current graph
axes	Create axes in arbitrary positions
subplot	Create axes in tiled positions
daspect	Data aspect ratio
pbaspect	Plot box aspect ratio
xlim	x limits
ylim	y limits
zlim	z limits

Viewpoint control

view	3-D graph viewpoint specification
viewmtx	View transformation matrix
rotate3d	Interactively rotate view of 3-D plot

Camera control

campos	Camera position
camtarget	Camera target
camva	Camera view angle
camup	Camera up vector
camproj	Camera projection

High-level camera control

camorbit	Orbit camera
campan	Pan camera
camdolly	Dolly camera
camzoom	Zoom camera
camroll	Roll camera
camlookat	Move camera and target to view specified objects
cameratoolbar	Interactively manipulate camera

High-level light control	
`camlight`	Creates or sets position of a light
`lightangle`	Spherical position of a light

Graph annotation	
`title`	Graph title
`xlabel`	x-axis label
`ylabel`	y-axis label
`zlabel`	z-axis label
`colorbar`	Display color bar (color scale)
`text`	Text annotation
`gtext`	Mouse placement of text
`plotedit`	Graph editing and annotation tools

Hard copy and printing	
`print`	Print graph or Simulink system; or save graph to M-file
`printopt`	Printer defaults
`orient`	Set paper orientation
`vrml`	Save graphics to VRML 2.0 file

15.15 Specialized graphs

`help specgraph`

Specialized 2-D graphs	
`area`	Filled area plot
`bar`	Bar graph
`barh`	Horizontal bar graph
`comet`	Comet-like trajectory
`compass`	Compass plot
`errorbar`	Error bar plot
`ezplot`	Easy-to-use function plotter
`ezpolar`	Easy-to-use polar coordinate plotter
`feather`	Feather plot

(continued on next page)

Specialized 2-D graphs (continued)

fill	Filled 2-D polygons
fplot	Plot function
hist	Histogram
pareto	Pareto chart
pie	Pie chart
plotmatrix	Scatter plot matrix
rose	Angle histogram plot
scatter	Scatter plot
stem	Discrete sequence or "stem" plot
stairs	Stairstep plot

Contour and 2½-D graphs

contour	Contour plot
contourf	Filled contour plot
contour3	3-D contour plot
clabel	Contour plot elevation labels
ezcontour	Easy-to-use contour plotter
ezcontourf	Easy-to-use filled contour plotter
pcolor	Pseudocolor (checkerboard) plot
voronoi	Voronoi diagram

Specialized 3-D graphs

bar3	3-D bar graph
bar3h	Horizontal 3-D bar graph
comet3	3-D comet-like trajectories
ezgraph3	General-purpose surface plotter
ezmesh	Easy-to-use 3-D mesh plotter
ezmeshc	Easy-to-use combination mesh/contour plotter
ezplot3	Easy-to-use 3-D parametric curve plotter
ezsurf	Easy-to-use 3-D colored surface plotter
ezsurfc	Easy-to-use combination surf/contour plotter

(continued on next page)

Specialized 3-D graphs (continued)

meshc	Combination mesh/contour plot
meshz	3-D mesh with curtain
pie3	3-D pie chart
ribbon	Draw 2-D lines as ribbons in 3-D
scatter3	3-D scatter plot
stem3	3-D stem plot
surfc	Combination surf/contour plot
trisurf	Triangular surface plot
trimesh	Triangular mesh plot
waterfall	Waterfall plot

Volume and vector visualization

vissuite	Visualization suite
isosurface	Isosurface extractor
isonormals	Isosurface normals
isocaps	Isosurface end caps
isocolors	Isosurface and patch colors
contourslice	Contours in slice planes
slice	Volumetric slice plot
streamline	Streamlines from 2-D or 3-D vector data
stream3	3-D streamlines
stream2	2-D streamlines
quiver3	3-D quiver plot
quiver	2-D quiver plot
divergence	Divergence of a vector field
curl	Curl and angular velocity of vector field
coneplot	3-D cone plot
streamtube	3-D stream tube
streamribbon	3-D stream ribbon
streamslice	Streamlines in slice planes
streamparticles	Display stream particles
interpstreamspeed	Interpolate streamline vertices from speed

(continued on next page)

Volume and vector visualization (continued)

subvolume	Extract subset of volume dataset
reducevolume	Reduce volume dataset
volumebounds	Returns x,y,z and color limits for volume data
smooth3	Smooth 3-D data
reducepatch	Reduce number of patch faces
shrinkfaces	Reduce size of patch faces

Image display and file I/O

image	Display image
imagesc	Scale data and display as image
colormap	Color look-up table
gray	Linear grayscale color map
contrast	Grayscale color map to enhance image contrast
brighten	Brighten or darken color map
colorbar	Display color bar (color scale)
imread	Read image from graphics file
imwrite	Write image to graphics file
imfinfo	Information about graphics file

Movies and animation

capture	Screen capture of current figure
moviein	Initialize movie frame memory
getframe	Get movie frame
movie	Play recorded movie frames
rotate	Rotate object about specified orgin and direction
frame2im	Convert movie frame to indexed image
im2frame	Convert index image into movie format

Color-related functions	
spinmap	Spin color map
rgbplot	Plot color map
colstyle	Parse color and style from string
ind2rgb	Convert indexed image to RGB image

Solid modeling	
cylinder	Generate cylinder
sphere	Generate sphere
ellipsoid	Generate ellipsoid
patch	Create patch
surf2patch	Convert surface data to patch data

15.16 Handle Graphics

help graphics

Figure window creation and control	
figure	Create figure window
gcf	Get handle to current figure
clf	Clear current figure
shg	Show graph window
close	Close figure
refresh	Refresh figure
openfig	Open new or raise copy of saved figure

Axis creation and control	
subplot	Create axes in tiled positions
axes	Create axes in arbitrary positions
gca	Get handle to current axes
cla	Clear current axes
axis	Control axis scaling and appearance
box	Axis box
caxis	Control pseudocolor axis scaling
hold	Hold current graph
ishold	Return hold state

Handle Graphics objects

`figure`	Create figure window
`axes`	Create axes
`line`	Create line
`text`	Create text
`patch`	Create patch
`rectangle`	Create rectangle, rounded rectangle, or ellipse
`surface`	Create surface
`image`	Create image
`light`	Create light
`uicontrol`	Create user interface control
`uimenu`	Create user interface menu
`uicontextmenu`	Create user interface context menu

Handle Graphics operations

`set`	Set object properties
`get`	Get object properties
`reset`	Reset object properties
`delete`	Delete object
`gco`	Get handle to current object
`gcbo`	Get handle to current callback object
`gcbf`	Get handle to current callback figure
`drawnow`	Flush pending graphics events
`findobj`	Find objects with specified property values
`copyobj`	Make copy of graphics object and its children
`isappdata`	Check if application-defined data exists
`getappdata`	Get value of application-defined data
`setappdata`	Set application-defined data
`rmappdata`	Remove application-defined data

Hard copy and printing	
`print`	Print graph or Simulink system; or save graph to M-file
`printopt`	Printer defaults
`orient`	Set paper orientation

Utilities	
`closereq`	Figure close request function
`newplot`	M-file preamble for NextPlot property
`ishandle`	True for graphics handles

ActiveX client functions (PC only)	
`actxcontrol`	Create an ActiveX control
`actxserver`	Create an ActiveX server

15.17 Graphical user interface tools

`help uitools`

GUI functions	
`uicontrol`	Create user interface control
`uimenu`	Create user interface menu
`ginput`	Graphical input from mouse
`dragrect`	Drag XOR rectangles with mouse
`rbbox`	Rubberband box
`selectmoveresize`	Interactively select, move, resize, or copy objects
`waitforbuttonpress`	Wait for key/buttonpress over figure
`waitfor`	Block execution and wait for event
`uiwait`	Block execution and wait for resume
`uiresume`	Resume execution of blocked M-file
`uistack`	Control stacking order of objects
`uisuspend`	Suspend the interactive state of a figure
`uirestore`	Restore the interactive state of a figure

GUI design tools

guide	Design GUI
inspect	Inspect object properties
align	Align uicontrols and axes
propedit	Edit property

Dialog boxes

axlimdlg	Axes limits dialog box
dialog	Create dialog figure
errordlg	Error dialog box
helpdlg	Help dialog box
imageview	Show image in figure with zoom
inputdlg	Input dialog box
listdlg	List selection dialog box
menu	Generate menu of choices for user input
movieview	Show movie in figure with replay button
msgbox	Message box
pagedlg	Page position dialog box
pagesetupdlg	Page setup dialog
printdlg	Print dialog box
printpreview	Display preview of figure to be printed
questdlg	Question dialog box
uigetpref	Question dialog box with preference support
soundview	Show sound in figure and play
uigetfile	Standard open file dialog box
uiputfile	Standard save file dialog box
uisetcolor	Color selection dialog box
uisetfont	Font selection dialog box
uiopen	Show open file dialog and call open on result
uisave	Show open file dialog and call save on result

(continued on next page)

Dialog boxes (continued)

uiload	Show open file dialog and call `load` on result
uiimport	Start the GUI for importing data (Import Wizard)
waitbar	Display wait bar
warndlg	Warning dialog box

Menu utilities

makemenu	Create menu structure
menubar	Computer-dependent default setting for MenuBar property
umtoggle	Toggle checked status of `uimenu` object
winmenu	Create submenu for Window menu item

Toolbar button group utilities

btngroup	Create toolbar button group
btnresize	Resize button group
btnstate	Query state of toolbar button group
btnpress	Button press manager for toolbar button group
btndown	Depress button in toolbar button group
btnup	Raise button in toolbar button group

Preferences

addpref	Add preference
getpref	Get preference
rmpref	Remove preference
setpref	Set preference
ispref	Test for existence of preference

Miscellaneous utilities	
allchild	Get all object children
clipboard	Copy and paste strings to and from system clipboard
edtext	Interactive editing of axes text objects
findall	Find all objects
findfigs	Find figures positioned off screen
getptr	Get figure pointer
getstatus	Get status text string in figure
hidegui	Hide/unhide GUI
listfonts	Get list of available system fonts in cell array
movegui	Move GUI to specified part of screen
guihandles	Return a structure of handles
guidata	Store or retrieve application data
overobj	Get handle of object the pointer is over
popupstr	Get popup menu selection string
remapfig	Transform figure objects' positions
setptr	Set figure pointer
setstatus	Set status text string in figure
uiclearmode	Clears the currently active interactive mode

15.18 Character strings

help strfun

General	
char	Create character array (string)
double	Convert string to numeric character codes
cellstr	Create cell array of strings from character array
blanks	String of blanks
deblank	Remove trailing blanks
eval	Execute string as a MATLAB expression

String tests

`ischar`	True for character array (string)
`iscellstr`	True for cell array of strings
`isletter`	True for letters of the alphabet
`isspace`	True for white space characters

String operations

`strcat`	Concatenate strings
`strvcat`	Vertically concatenate strings
`strcmp`	Compare strings
`strncmp`	Compare first N characters of strings
`strcmpi`	Compare strings ignoring case
`strncmpi`	Compare first N characters of strings ignoring case
`findstr`	Find one string within another
`strfind`	Find one string within another
`strjust`	Justify character array
`strmatch`	Find possible matches for string
`strrep`	Replace string with another
`strtok`	Find token in string
`upper`	Convert string to uppercase
`lower`	Convert string to lowercase

String to number conversion

`num2str`	Convert number to string
`int2str`	Convert integer to string
`mat2str`	Convert matrix to `eval`'able string
`str2double`	Convert string to double-precision value
`str2num`	Convert string matrix to numeric array
`sprintf`	Write formatted data to string
`sscanf`	Read string under format control

Base number conversion	
hex2num	Convert IEEE hexadecimal to double-precision number
hex2dec	Convert hexadecimal string to decimal integer
dec2hex	Convert decimal integer to hexadecimal string
bin2dec	Convert binary string to decimal integer
dec2bin	Convert decimal integer to binary string
base2dec	Convert base B string to decimal integer
dec2base	Convert decimal integer to base B string

15.19 File input/output

help iofun

File import/export functions	
dlmread	Read delimited text file
dlmwrite	Write delimited text file
load	Load workspace from MATLAB (.mat) file
importdata	Load workspace variables disk file
wk1read	Read spreadsheet (WK1) file
wk1write	Write spreadsheet (WK1) file
xlsread	Read spreadsheet (XLS) file

Image file import/export	
imfinfo	Return information about graphics file
imread	Read image from graphics file
imwrite	Write image to graphics file

Audio file import/export	
auread	Read NeXT/SUN (.au) sound file
auwrite	Write NeXT/SUN sound file
wavread	Read Microsoft WAVE (.wav) sound file
wavwrite	Write Microsoft WAVE sound file

Video file import/export

`aviread`	Read movie (AVI) file
`aviinfo`	Return information about AVI file
`avifile`	Create a new AVI file
`movie2avi`	Create AVI movie from MATLAB movie

Formatted file I/O

`fgetl`	Read line from file, discard newline character
`fgets`	Read line from file, keep newline char.
`fprintf`	Write formatted data to file
`fscanf`	Read formatted data from file
`input`	Prompt for user input
`textread`	Read formatted data from text file

String conversion

`sprintf`	Write formatted data to string
`sscanf`	Read string under format control
`strread`	Read formatted data from text string

File opening and closing

`fopen`	Open file
`fclose`	Close file

Binary file I/O

`fread`	Read binary data from file
`fwrite`	Write binary data to file

File positioning

`feof`	Test for end-of-file
`ferror`	Inquire file error status
`frewind`	Rewind file
`fseek`	Set file position indicator
`ftell`	Get file position indicator

File name handling

fileparts	Filename parts
filesep	Directory separator for this platform
fullfile	Build full filename from parts
matlabroot	Root directory of MATLAB installation
mexext	MEX filename extension for this platform
partialpath	Partial pathnames
pathsep	Path separator for this platform
prefdir	Preference directory name
tempdir	Get temporary directory
tempname	Get temporary file

HDF library interface help

hdf	MEX-file interface to the HDF library
hdfan	HDF multifile annotation interface
hdfdf24	HDF raster image interface
hdfdfr8	HDF 8-bit raster image interface
hdfh	HDF H interface
hdfhd	HDF HD interface
hdfhe	HDF HE interface
hdfml	MATLAB-HDF gateway utilities
hdfsd	HDF multifile scientific dataset interface
hdfv	HDF V (Vgroup) interface
hdfvf	HDF VF (Vdata) interface
hdfvh	HDF VH (Vdata) interface
hdfvs	HDF VS (Vdata) interface

HDF-EOS library interface help

hdfgd	HDF-EOS grid interface
hdfpt	HDF-EOS point interface
hdfsw	HDF-EOS swath interface

Serial port support

serial	Construct serial port object

Command window I/O	
`clc`	Clear Command window
`disp`	Display array
`home`	Send cursor home
`input`	Prompt for user input
`pause`	Wait for user response

FIG file support for plotedit and printframes	
`hgload`	Load Handle Graphics object from a file
`hgsave`	Saves an HG object heirarchy to a file

Utilities	
`str2rng`	Convert spreadsheet range string to numeric array
`wk1const`	WK1 record type definitions
`wk1wrec`	Write a WK1 record header

15.20 Time and dates

`help timefun`

Current date and time	
`now`	Current date and time as date number
`date`	Current date as date string
`clock`	Current date and time as date vector

Basic functions	
`datenum`	Serial date number
`datestr`	String representation of date
`datevec`	Date components

Date functions	
`calendar`	Calendar
`weekday`	Day of week
`eomday`	End of month
`datetick`	Date formatted tick labels

Timing functions	
cputime	CPU time in seconds
tic	Start stopwatch timer
toc	Stop stopwatch timer
etime	Elapsed time
pause	Wait in seconds

15.21 Data types and structures

help datatypes

Data types (classes)	
double	Convert to double precision
sparse	Create sparse matrix
char	Create character array (string)
cell	Create cell array
struct	Create or convert to structure array
single	Convert to single precision
uint8	Convert to unsigned 8-bit integer
uint16	Convert to unsigned 16-bit integer
uint32	Convert to unsigned 32-bit integer
int8	Convert to signed 8-bit integer
int16	Convert to signed 16-bit integer
int32	Convert to signed 32-bit integer
inline	Construct inline object
function_handle	Function handle array
javaArray	Construct a Java array
javaMethod	Invoke a Java method
javaObject	Invoke a Java object constructor

Multidimensional array functions	
cat	Concatenate arrays
ndims	Number of dimensions
ndgrid	Generate arrays for N-D functions and interpolation

(continued on next page)

Multidimensional array functions (continued)

permute	Permute array dimensions
ipermute	Inverse permute array dimensions
shiftdim	Shift dimensions
squeeze	Remove singleton dimensions

Cell array functions

cell	Create cell array
cellfun	Functions on cell array contents
celldisp	Display cell array contents
cellplot	Display graphical depiction of cell array
num2cell	Convert numeric array into cell array
deal	Deal inputs to outputs
cell2struct	Convert cell array into structure array
struct2cell	Convert structure array into cell array
iscell	True for cell array

Structure functions

struct	Create or convert to structure array
fieldnames	Get structure field names
getfield	Get structure field contents
setfield	Set structure field contents
rmfield	Remove structure field
isfield	True if field is in structure array
isstruct	True for structures

Function handle functions

@	Create function_handle
func2str	Convert function_handle array into string
str2func	Convert string into function_handle array
functions	List functions associated with a function_handle

Object-oriented programming functions

class	Create object or return object class
struct	Convert object to structure array
methods	List names and properties of class methods
methodsview	View names and properties of class methods
isa	True if object is a given class
isjava	True for Java objects
isobject	True for MATLAB objects
inferiorto	Inferior class relationship
superiorto	Superior class relationship
substruct	Create structure argument for subsref/subsasgn

Overloadable operators

minus	Overloadable method for a-b
plus	Overloadable method for a+b
times	Overloadable method for a.*b
mtimes	Overloadable method for a*b
mldivide	Overloadable method for a\b
mrdivide	Overloadable method for a/b
rdivide	Overloadable method for a./b
ldivide	Overloadable method for a.\b
power	Overloadable method for a.^b
mpower	Overloadable method for a^b
uminus	Overloadable method for -a
uplus	Overloadable method for +a
horzcat	Overloadable method for [a b]
vertcat	Overloadable method for [a;b]
le	Overloadable method for a<=b
lt	Overloadable method for a<b
gt	Overloadable method for a>b
ge	Overloadable method for a>=b

(continued on next page)

Overloadable operators (continued)	
eq	Overloadable method for a==b
ne	Overloadable method for a~=b
not	Overloadable method for ~a
and	Overloadable method for a&b
or	Overloadable method for a|b
subsasgn	Overloadable method for a(i)=b, a{i}=b, and a.field=b
subsref	Overloadable method for a(i), a{i}, and a.field
colon	Overloadable method for a:b
end	Overloadable method for a(end)
transpose	Overloadable method for a.'
ctranspose	Overloadable method for a'
subsindex	Overloadable method for x(a)
loadobj	Called to load object from .mat file
saveobj	Called to save object to .mat file

15.22 Version control commands

`help verctrl`

Checkin/checkout	
checkin	checkin files to version control system
checkout	checkout files
undocheckout	undo checkout files

Specific version control	
rcs	Version control actions using RCS
pvcs	Version control actions using PVCS
clearcase	Version control actions using ClearCase
sourcesafe	Version control using Visual SourceSafe
customverctrl	Custom version control template

15.23 Microsoft Windows functions

`help winfun`

ActiveX client functions	
`actxcontrol`	Create an ActiveX control
`actxserver`	Create an ActiveX server
`winfun\activex`	ActiveX class

ActiveX demos	
`mwsamp`	Sample ActiveX control creation
`sampev`	Sample event handler for ActiveX server

DDE client functions	
`ddeadv`	Set up advisory link
`ddeexec`	Send string for execution
`ddeinit`	Initiate DDE conversation
`ddepoke`	Send data to application
`ddereq`	Request data from application
`ddeterm`	Terminate DDE conversation
`ddeunadv`	Release advisory link

15.24 Demos

Type `help demos` to see the list of MATLAB demos.
Section 15.26 lists the Symbolic Math Toolbox demos.

15.25 Preferences

`help local`

Saved preferences files	
`startup`	User startup M-file
`finish`	User finish M-file
`matlabrc`	Master startup M-file
`pathdef`	Search path defaults
`docopt`	Web browser defaults
`printopt`	Printer defaults

Preference commands	
cedit	Set command line editor keys
terminal	Set graphics terminal type

Configuration information	
hostid	MATLAB server host ID number
license	License number
version	MATLAB version number

15.26 Symbolic Math Toolbox

help symbolic

Calculus	
diff	Differentiate
int	Integrate
limit	Limit
taylor	Taylor series
jacobian	Jacobian matrix
symsum	Summation of series

Linear algebra	
diag	Create or extract diagonals
triu	Upper triangle
tril	Lower triangle
inv	Matrix inverse
det	Determinant
rank	Rank
rref	Reduced row echelon form
null	Basis for null space
colspace	Basis for column space
eig	Eigenvalues and eigenvectors
svd	Singular values and singular vectors
jordan	Jordan canonical (normal) form
poly	Characteristic polynomial
expm	Matrix exponential

Simplification

simplify	Simplify
expand	Expand
factor	Factor
collect	Collect
simple	Search for shortest form
numden	Numerator and denominator
horner	Nested polynomial representation
subexpr	Rewrite in terms of subexpressions
subs	Symbolic substitution

Solution of equations

solve	Symbolic solution of algebraic equations
dsolve	Symbolic solution of differential equations
finverse	Functional inverse
compose	Functional composition

Variable precision arithmetic

vpa	Variable precision arithmetic
digits	Set variable precision accuracy

Integral transforms

fourier	Fourier transform
laplace	Laplace transform
ztrans	Z transform
ifourier	Inverse Fourier transform
ilaplace	Inverse Laplace transform
iztrans	Inverse Z transform

Conversions

`double`	Convert symbolic matrix to double
`poly2sym`	Coefficient vector to symbolic polynomial
`sym2poly`	Symbolic polynomial to coefficient vector
`char`	Convert sym object to string

Basic operations

`sym`	Create symbolic object
`syms`	Shortcut for constructing symbolic objects
`findsym`	Determine symbolic variables
`pretty`	Pretty print a symbolic expression
`latex`	LaTeX representation of a symbolic expression
`ccode`	C code representation of a symbolic expression
`fortran`	FORTRAN representation of a symbolic expression

Special functions

`sinint`	Sine integral
`cosint`	Cosine integral
`zeta`	Riemann zeta function
`lambertw`	Lambert W function

String handling utilities

`isvarname`	Check for a valid variable name (MATLAB toolbox)
`vectorize`	Vectorize a symbolic expression

Pedagogical and graphical applications

`rsums`	Riemann sums
`ezcontour`	Easy-to-use contour plotter
`ezcontourf`	Easy-to-use filled contour plotter
`ezmesh`	Easy-to-use mesh (surface) plotter
`ezmeshc`	Easy-to-use mesh/contour plotter
`ezplot`	Easy-to-use function implicit and parametric curve plotter
`ezplot3`	Easy-to-use spatial curve plotter
`ezpolar`	Easy-to-use polar coordinates plotter
`ezsurf`	Easy-to-use surface plotter
`ezsurfc`	Easy-to-use surface/contour plotter
`funtool`	Function calculator
`taylortool`	Taylor series calculator

Demonstrations

`symintro`	Introduction to the Symbolic Math Toolbox
`symcalcdemo`	Calculus demonstration
`symlindemo`	Demonstrate symbolic linear algebra
`symvpademo`	Demonstrate variable precision arithmetic
`symrotdemo`	Study plane rotations
`symeqndemo`	Demonstrate symbolic equation solving

Access to Maple (not in Student Version)

`maple`	Access Maple kernel
`mfun`	Numeric evaluation of Maple functions
`mfunlist`	List of functions for MFUN
`mhelp`	Maple help
`procread`	Install a Maple procedure

16. Additional Resources

The MathWorks, Inc., and others provide a wide range of products that extend MATLAB's capabilities. Some are collections of M-files called toolboxes. One of these has already been introduced (the Symbolic Math Toolbox). Also available is Simulink, an interactive graphical system for modeling and simulating dynamic nonlinear systems. The `ver` command lists the toolboxes and Simulink components included in your installation. These can be explored via the command `help` or from the Launch Pad window. Similar to MATLAB toolboxes, Simulink has domain-specific add-ons called blocksets.

16.1 MATLAB

MATLAB Compiler (convert M-files to C/C++)
MATLAB C/C++ Math Library
MATLAB C/C++ Graphics Library
MATLAB Report Generator
MATLAB Runtime Server
MATLAB Web Server

16.2 MATLAB toolboxes

Math and Analysis Toolboxes:
Optimization
Statistics
Neural Network
Symbolic/Extended Symbolic Math
Partial Differential Equation
Mapping (geographic information)
Spline

Data Acquisition and Import Toolboxes:
Data Acquisition

Instrument Control
Database
Excel Link

Signal & Image Processing Toolboxes:
Signal Processing
Image Processing
Communications
System Identification
Wavelet
Filter Design
Motorola DSP Developer's Kit
Developer's Kit for Texas Instruments DSP

Control Design Toolboxes:
Control System
Fuzzy Logic
Robust Control
μ-Analysis and Synthesis
LMI (linear matrix inequality) Control
Model Predictive Control

Finance and Economics Toolboxes:
Financial
Financial Time Series
GARCH (volatility analysis)
Financial Derivatives
Datafeed (connect to financial data providers)

16.3 Simulink

Simulink Performance Tools
Stateflow
Stateflow Coder
Real-Time Windows Target

Real-Time Workshop
Real-Time Workshop Embedded Coder
Real-Time Workshop Ada Coder
xPC Target
xPC Target Embedded Option
Simulink Report Generator
Requirements Management Interface

16.4 Simulink blocksets

CDMA Reference (mobile phone simulation)
Communications
Dials & Gauges
DSP (Digital Signal Processing)
Fixed-Point
Nonlinear Control Design
Power System

Index